FM 21-50

US ARMY FIELD MANUAL

RANGER TRAINING AND RANGER OPERATIONS

1962

CIVILIAN REFERENCE EDITION

UNABRIDGED HISTORIC MILITARY HANDBOOK ON TRAINING PROGRAMS FOR INTENSIVE OPERATIONS, WILDERNESS PROCEDURES, AND MOUNTAINEERING

U.S. DEPARTMENT OF THE ARMY

Doublebit Press

www.doublebitpress.com | Cherry, IL, USA

Original content under the public domain; unrestricted for civilian distribution.
Originally published in 1962 by the U.S. Department of the Army.

A Part of the Military Outdoors Skills Series

Doublebit Press Civilian Reference Edition ISBNs
Hardcover: 978-1643892108
Paperback: 978-1643892115

FIELD MANUAL No. 21-50

HEADQUARTERS, DEPARTMENT OF THE ARMY
WASHINGTON 25, D.C., *23 Januaay 1962*

RANGER TRAINING AND RANGER OPERATIONS

		Paragraphs	Page
CHAPTER 1.	INTRODUCTION	1-3	3
2.	THE RANGER	4-6	4
3.	THE TRAINING PROGRAM		
Section I.	Preparation for a Ranger training program	7-9	7
II.	Conduct of the Ranger training program	10-30	14
CHAPTER 4.	RANGER OPERATIONS		
Section I.	Introduction	31	57
II.	Airmobile operations	32-37	57
III.	Ambush and roadblock operations	38-41	63
IV.	Cliff assault operations	42-45	69
V.	Extended operations	46-57	80
VI.	Small unit waterborne operations	58-60	114
VII.	Antiguerrilla operations	61-66	130
APPENDIX I.	REFERENCES		150
II.	TRAINING SCHEDULES		152
III.	LESSON SUBJECT OUTLINES		160
IV.	CONDUCT OF PROBLEMS REFERENCED IN SCHEDULING		208
V.	EXAMPLE MOUNTAINEERING LESSON OUTLINES		233
VI.	EXAMPLE SURVIVAL LESSON OUTLINE		268
VII.	PATROL TIPS		278
VIII.	GENERAL CONSIDERATIONS OF PATROLLING		285
IX.	EXAMPLE PROBLEM OUTLINE TO A TRAINING MEMORANDUM		291
X.	PROBLEM COMMUNICATIONS SUPPORT		295
XI.	GENERAL CRITIQUE CHECKLIST		298
XII.	EXAMPLE PATROL ORDER (AMBUSH PATROL)		305
XIII.	EXAMPLE PATROL ORDER (GUERRILLA RAID)		309
XIV.	EXAMPLE LANE INSTRUCTOR'S INSTRUCTIONS (GUERRILLA RAID)		314
XV.	EXAMPLE COMMAND POST INSTRUCTIONS		318
XVI.	EXAMPLE OUTPOST INSTRUCTIONS		320
XVII.	RANGER HISTORY		321
GLOSSARY			334
INDEX			337

***This manual supersedes FM 21-50, 20 August 1957.**

CHAPTER 1

INTRODUCTION

1. Purpose

This manual primarily is a guide for establishing a Ranger training program and conducting Ranger operations.

2. Scope

a. The Ranger training manual contains the necessary information, organization, doctrine, and general guidance that a commander needs to develop and initiate a Ranger training program. The material in this manual is applicable for training of all regular and special purpose units of the United States Army.

b. The material in this manual is applicable without modification to both nuclear and nonnuclear warfare.

c. Users of this manual are encouraged to submit recommended changes or comments to improve the manual. Comments should be keyed to the specific page, paragraph, and line of the text in which the change is recommended. Reasons should be provided for each comment to insure understanding and complete evaluation. Comments should be forwarded direct to Commandant, United States Army Infantry School, Fort Benning, Ga.

3. Publications

It will be necessary for units conducting this training to consult appropriate publications. Those references pertinent to Ranger training are found in appendix I.

CHAPTER 2

THE RANGER

4. The Ranger Imprint

The Ranger Imprint is: "Pride, confidence, self-determination, and the ability to lead, endure, and succeed regardless of the odds or obstacles of the enemy, weather, and terrain." Ranger training conducted by a unit must be prepared and presented with the Ranger Imprint as a major end product. To develop this the following factors and methods are incorporated to the maximum extent possible into all instruction:

a. The instructor sets the example.

b. Continuity of instruction.

c. Close contact between instructor, supervisory personnel, and Soldier.

d. Rugged, realistic training situations that test the individual's confidence and ability to overcome inherent fears.

e. Require the Soldier to exert and express himself.

f. Provide situations and opportunity to challenge the Soldier.

g. Motivate the Soldier to obtain the maximum instruction regardless of unfavorable conditions.

h. Development of sense of accomplishment and progress.

i. Development of individual skills and competitive spirit.

j. Provide opportunities for Soldiers to exercise initiative, leadership, and command abilities through frequent rotation of chain of command.

k. Progressive physical conditioning program continuous throughout the course.

l. Maximum efficiency in planning and conduct of instruction.

m. Development of an appreciation for the maintenance of equipment, supply economy, and its effect on the accomplishment of the mission.

n. Maximum night training.

5. The Ranger Concept

Ranger training is realistic, rough, and to a degree hazardous. It is designed to develop the individual's self-confidence, leadership, and skill in the application of basic principles and techniques. Ranger training will teach and train individuals to overcome mental and physical obstacles by using combat realistic situations in small unit tactical exercises. Emphasis is placed on teaching survival and land navigation principles and techniques; the development of good leadership habits through the use of a tactical vehicle—the patrol; developing physical and mental endurance. Ranger (battle) confidence is developed by placing the Soldier in a combat environment where he must learn to survive, move, and fight at extended distances behind enemy lines. He must do this with minimum support. Hunger, fatigue, and tactical realism will uncover strengths and weaknesses that an individual does not know he possesses. Through Ranger training, he gains an insight into himself and his fellow Soldier.

6. Training Phases

Ranger training is an extension of the Soldier's professional training. This training will instill the Ranger Imprint on *individuals* and develop exceptional *unit* capabilities. Although there is only one Ranger training phase, the commander must realize that he should work toward these goals during the preceding training programs, and that there is a definite review and conditioning period that preceded the Ranger training phase.

a. Pre-Ranger Training Phase. This is the period immediately preceding the Ranger training phase during which the Soldier is prepared physically, mentally, and emotionally for the advanced training to follow. The pre-Ranger phase is not included as part of the training outlined in this manual. Units about to undergo Ranger training should utilize this time in Ranger indoctrination, physical conditioning, reviews of map reading, use of the compass, patrolling, survival, and other small unit tactics, principles, and techniques. Confidence is developed in the Soldier by exacting physical and mental standards. The Ranger indoctrination period is of utmost importance because the initial standards for each Soldier will determine his reaction in the Ranger training phase. Rigid discipline is enforced. Strict adherence to details is mandatory. The attainment of the highest standards is demanded.

b. Ranger Training Phase. In this phase, Soldiers are forced into a variety of tactical situations. The patrol is used as the teaching vehicle. Patrols are planned and prepared, rehearsed, and then executed. Patrols will vary in size depending on the mission. Terrain should be varied and challenge navigational ability. Training must be conducted without regard to climatic conditions. All patrols are accompanied and closely observed by a qualified officer or noncommissioned officer. Aggressor action is used against patrols so that assigned leaders are required to make prompt, sound decisions and take immediate action. Leadership positions are rotated. The number of leader positions during any one exercise depends upon the size of the patrol, duration of the operation, and the mission. Upon completion of an exercise, the unit is given a detailed critique. All advanced instruction during this phase is presented in a realistic combat rear area atmosphere. The rear area environment sets the stage for the field training. Advanced subjects presented during this phase are discussed in paragraphs 7 through 9.

CHAPTER 3

THE TRAINING PROGRAM

Section I. PREPARATION FOR A RANGER TRAINING PROGRAM

7. General

Although a Ranger training program is applicable to individual training conducted by a higher headquarters, it is tailored to fit the squad, platoon, and larger units. The squad can provide the base for the patrol in this training. With the squad as the base, the Ranger training program can be expanded to meet the specialized requirements of larger units organized for special missions such as antiguerrilla operations, combat raids, extended patrol activities, etc. Preparation for this program varies slightly from that of other field training exercises. Generally, the variations are limited to equipment, instructor qualification, and terrain.

8. Training Subjects

The subjects discussed in this paragraph are mandatory for successful completion of the Ranger training program.

a. Physical Conditioning. Physical conditioning is an integral part of all training programs. In Ranger training, its importance is emphasized because of the requirements for strength, endurance, agility, and coordination. To be combat ready the Soldier must be technically, mentally, and physically fit. It is important that the unit conduct a vigorous conditioning program prior to starting the Ranger training phase. Progressive physical conditioning is continuous throughout the phase.

b. Combatives. Combatives include hand-to-hand combat and bayonet. These subjects are included in the program to instill aggressiveness and the will to win, develop self-confidence, and aid in the development of physical fitness.

c. Demolitions. Demolitions are used extensively in Ranger operations. Ranger demolitions are simple, easily prepared, and effective. Every Soldier must know how to prepare, calculate, and place these charges.

d. Escape and Evasion. Because the Ranger works primarily behind enemy lines, escape and evasion is a necessary part of his training. The Soldier must know how to evade the enemy and how to escape if captured.

e. Confidence Tests. Confidence tests are included in Ranger training to increase the confidence of the Soldier by requiring him to negotiate obstacles which appear more difficult than they actually are. Normally there are three confidence tests used: the rope drop, the suspension traverse, and the confidence (combat conditioning) course.

f. Patrol Planning, Orders, and Techniques. The patrol is the basic vehicle for Ranger training. The Soldier must be taught the basic principles and techniques of planning, preparing, and executing Ranger type patrols and missions. The patrol cannot be ignored by the small unit leader. He must master the organization, requirements, equipment, and support for patrols to be successful in planning and preparing for future larger unit actions.

g. Survival. The Soldier must be acquainted with methods of survival in all types of terrain and under all weather conditions. He must be able to identify edible plant and animal life, construct shelters, and survive by his own resources. Survival becomes an integral part of the Ranger training program because the Soldier should be required to supplement his diet by living off the land.

h. Orientation. The orientation is used to familiarize the Soldier with the training to be conducted, standards to be maintained, training procedures, the aggressor enemy, and the existing combat situation.

i. Land Navigation and Map Reading. A complete understanding of map and compass work is necessary in Ranger training. Map and aerial photograph reading will be a review for the Soldier. The ability to land navigate is essential because the greatest percentage of Ranger training is fieldwork.

j. Intelligence. The importance of intelligence cannot be overemphasized. The Soldier must be familiar with collecting, recording, and forwarding information of intelligence value.

k. Combat Formations. The tactical advantage of combat formations must be stressed in training. The use of common signals must be understood by all Soldiers.

l. Inspections. Inspections create high standards of discipline, appearance, and maintenance of equipment.

m. Ambush and Roadblock Techniques. Ambush and roadblock techniques are taught because of extensive use in Ranger operations.

n. Aerial Resupply and Airmobile Operations. Instruction in resupply, evacuation, and air movement is important in a Ranger training program. Many Ranger type missions will utilize aircraft in support of the operation.

o. Cliff Assault Techniques. This subject is included in the program to familiarize the unit with techniques of scaling and assaulting cliffs, landing on beaches, and security and organization of a raiding party.

p. River Crossing Expedients. The Soldier must know how to cross rivers, streams, and small bodies of water effectively to accomplish his mission.

q. Mountain Techniques and Expedients. This is included in Ranger training to familiarize the Soldier with basic military mountaineering techniques, equipment, and expedients. These provide additional techniques for use in accomplishing the mission.

r. Summary. The previously mentioned subjects are considered of general importance in a Ranger training program. (For details of these subjects, refer to app. III.) The local commander should not consider himself limited to these subjects. Prevailing conditions of climate and terrain may effectively dictate inclusion of other material.

9. Development of the Program

In order to develop the training program, a commander must understand the Ranger Imprint and the concept of Ranger training. This training program must be developed realistically with the combat situations serving as a basis for all planning. Initial planning for the commander is outlined in this paragraph.

a. Training Schedules. Schedules included in this manual have received troop tests (nonvolunteer) and are based on instruction presented to regular Ranger course students. Subjects and exercises are arranged so that training may be conducted in a logical, realistic, and efficient manner. Necessary briefings and tactical training help to maintain the atmosphere of realism throughout the training. Ranger training encompasses many subjects not included in this manual. Commanders should consider expanding their training to include subjects and missions which may be encountered on future battlefields. In addition, desert, arctic, amphibious, airborne, cliff assault, and jungle operations should be considered. Training schedules are found in appendix II.

b. Lesson Subject Outlines. Subject outlines are included in this manual as guides to reduce preparation time. Lesson subject outlines are listed in appendix III.

c. Selection of Personnel.

(1) *Officer in charge.* The officer in charge of the training program should be carefully selected since his initiative and supervision to a great degree determine its success. Ranger qualification is a highly desirable prerequisite for this position.

(2) *Requirements.* The following requirements are included as a guide for the training of company-size units. Requirements can be varied to meet situations which indicate that an augmentation of personnel is desired. If possible, all members of the training cadre should be Ranger qualified.

(*a*) *Planning personnel.*

1. One officer thoroughly familiar with current directives, training status of units, and training of personnel. This officer will determine which subjects to schedule as well as prepare the administrative plan for the course.

2. One officer to train instructors, prepare problems, and to lay out training areas.

(*b*) *Instructor personnel.*

1. One instructor is needed for each squad under this training program. The number of cadre officers and enlisted instructors is determined by the number of squads and platoons trained; i.e., one cadre noncommissioned officer per squad, one cadre officer per platoon. These are minimum requirements. Planning and supervisory personnel can be included as instructors; one instructor will also function as the operations and training officer for the course.

2. One officer and enlisted assistant to act as the course tactical officer and enlisted assistant. Primary functions of these personnel are to supervise, maintain records, counsel, conduct inspections, and evaluate the overall performance of those undergoing training. Extremely valuable information can be provided to the commander relative to the promotion potential, knowledge, and physical capabilities of individuals and units. Only experienced officers and noncommissioned officers should be selected as tactical officers or tactical noncommissioned officers.

(3) *Class organization.*

(*a*) *Individual training.* A chain of command, to include a company commander, executive officer, first sergeant, platoon leaders, platoon sergeants, and squad leaders operates directly under the tactical officer. The class is divided into officer and enlisted platoons. Classes are separated into groups by rank; normally these are officers, senior noncommissioned, and junior noncommissioned officer and potential noncommissioned officer groups. Elimination of rank within groups is desirable for effective rotation of command during patrols as well as for administrative purposes.

(*b*) *Unit training.* Maintain unit integrity.

(4) *Aggressor troops.* When possible, the number of aggressor troops should be on a one for one basis. Aggressor actions, to a great extent, determine the realism of tactical problems. If there is a limited aggressor capability, dispersion over a

wide area requires them to perform several missions during each phase.

(5) *Modifications.* If there is no aggressor support available, problems may be modified into a two-sided maneuver. Experts in certain subjects may be found within the company. The commander or training officer must be careful not to lose the momentum of his training program or the quality of his instruction through this modification.

d. Planning Guidance. The officer in charge, regardless of the level of the training course to be conducted, must receive guidance from his commander or higher headquarters. This guidance should include—

(1) The training program outline. This may be in detail and include the scope of each problem to be conducted (app. IX).

(2) Authority to obtain equipment and information as to its availability.

(3) Available training areas and facilities.

(4) Source of personnel to be trained.

(5) Support agencies and aggressor troops.

e. Training Areas and Facilities.

(1) *Planning.* The officer in charge of the Ranger training program should familiarize himself with the garrison and field training exercises prior to establishing the training locations. In some areas, training ranges must be requested well in advance. For this reason, the planning officer must be familiar with problem requirements in order to secure as many varied training areas for reconnaissance as is possible. See figure 1.

(2) *Reconnaissance.* In reconnoitering the proposed training sites, the planning officer should locate a desirable area of sufficient size to conduct the maximum number of problems. It should be a centralized bivouac site without loss of realism and with a maximum potential for creating a tactical situation.

(*a*) *Map reconnaissance.* Study a map 1:50,000 *or smaller* of the problem training areas. The capabilities and limitations of the terrain as to contour, vegetation, streamlines, water basins, access routes and trails, etc., are studied. Several locations of the base camp are visualized; several mental "run-throughs" of the patrol problems are made; and tentative locations of the problem sites are selected.

(*b*) *Other reconnaissance.* Combined air-ground reconnaissance is the most desirable method of inspecting the areas. The officer in charge should inspect logical objectives of the problems. After landing, a detailed ground reconnaissance should be completed quickly. From the air, the officer in charge can quickly and effectively trace desirable patrol routes. He can spot logical friendly and enemy outposts, battle dispositions, obstacles, routes of communication, and observation posts. Landing at a selected base campsite, a detailed ground reconnaissance can be completed by walking the entire area visualizing the camp layout, physical conditioning area, running area, etc. Upon completion of this reconnaissance, a decision on the training area can be made. Utmost isolation is desired. Local civilian activity detracts from the realism and should be curtailed in the training area.

f. Specific Training Areas. Suitable training areas are necessary for effective training. Most classroom subjects should be taught prior to movement to the field.

(1) *Bivouac.* The bivouac should contain areas suitable to the conduct of informal classroom training. It should contain at least one all-weather running road.

(*a*) *Formal class area.* A field classroom may be constructed using sandbags or logs for seating arrangements. When appropriate, a semicircular, sloping bank, or mound, on the fringe of the bivouac area presents the most advantageous classroom area for conducting classes, patrol briefings, and debriefings.

(*b*) *Hand-to-hand combat.* A circular, sawdust pit is best suited for a hand-to-hand combat training area. For a company-size class the sawdust pit should be 21.34 meters in diameter, the innermost 3.05 meters comprising an instructor's demonstration platform. For each additional 100 Soldiers, the radius is extended an additional 9.14 meters. This provides approximately 3.35 to 3.62 square meters of training space per Soldier. The sawdust should be at least 15¼ cm (6 ins.) deep. In the event sawdust is not available the earth should be plowed, harrowed, and finely raked to obtain a suitable surface.

(*c*) *Compass course.* The night compass course should have control roads and trails running through and around the area. The area should be bounded by prominent terrain features and include the same type of terrain and vegetation as the patrol areas. A minimum of open areas will prevent Soldiers from running into and observing the paths of others. Isolation of buddy teams on the course is desired.

(2) *Mountains.* This terrain is excellent for the conduct of patrols and offers the maximum natural training aids for con-

duct of mountaineering techniques. Desirable features of mountainous areas are—

(*a*) Sheer cliffs, 7 to 15½ meters high for rappelling and cliff assault technique instruction. There should be a good platform on the top of the cliff and a large work area at the base of the cliff. Access routes should be convenient to the top of the cliff.

(*b*) Ravines and streams provide suitable obstacles for rope bridges, suspension traverse, and other rope techniques. Heavy timber provides anchor points for fixing ropes. Access to the area should be convenient in the event emergency evacuation is necessary.

(*c*) Slopes require mastering of mountain walking on varied surfaces. Wooded, tufted, and rocky slopes require different techniques.

(3) *Jungle and swamp.* While suitable jungle training areas are not normally available, suitable swampy areas may be found. Some of these swampy areas are covered with tough, dense thickets and approach jungle vegetation as barriers to travel.

(4) *Desert.* Suitable desert training areas are located in Western and Southwestern United States.

(5) *Northern.* Portions of the Northern areas are heavily vegetated and mountainous which lend themselves to Ranger type training activities. In addition, the capability to undertake cold weather operations is a distinct advantage to a unit. Many of the same training techniques are applicable to operations in both warm and cold climates.

g. Coordination.

(1) Upon completion of the reconnaissance and final selection of the training areas, terrain requests should be submitted in sufficient time to allow for detailed planning by appropriate headquarters.

(2) Initial coordination is made to insure that sufficient vehicles, personnel, and expendable items are available for the program.

(3) Final coordination is made after positive selection of the training areas. This coordination is made in the following fields:

(*a*) Personnel.

(*b*) Time scheduling.

(*c*) Transportation.

(*d*) Communications.

(*e*) Control.

(*f*) Rations.

(*g*) Equipment.

(*h*) Ammunition and demolitions.

(*i*) Civilian liaison (for training on public or private land).

(*j*) Safety.

(*k*) Rehearsal.

(*l*) Completion date.

h. Rehearsal. The officer in charge supervises a rehearsal of the cadre in the conduct of the training program. Rehearsals should be conducted well in advance of the scheduled exercise to permit the correction of errors and changes required to improve training. The commander who directed the training exercise should be present at the rehearsal to critique and offer suggestions for improvement. In cases when neither time nor troops will be available to conduct a full rehearsal, instruction which has the greatest effect on the Soldier is rehearsed to perfection.

i. Summary.

(1) The Ranger training program is clearly divided into two echelons of responsibility; that which is held by the initiating headquarters and those responsibilities delegated to the officer in charge of the training program.

(2) The initiating headquarters maintains responsibility for—

(*a*) Selection of the officer in charge.

(*b*) Determining the scope, concept, and type of Ranger training to be conducted.

(*c*) Establishing effective training dates and designating individuals or units to undergo training.

(*d*) Outlining support and requesting procedures to include approximate availability of troops and equipment.

(3) The officer in charge generally is made responsible for the detailed planning to include requirements and overall conduct of the program. He must, based on guidance from the initiating headquarters—

(*a*) Select and request training areas and facilities.

(*b*) Request troops, cadre, and equipment.

(*c*) Prepare detailed training schedules, instruction lesson plans, and problem vault files.

(*d*) Rehearse and train instructional cadre.

(*e*) Establish necessary Soldier evaluation systems.

(*f*) Effect all coordination necessary prior to and during the training phase.

Section II. CONDUCT OF THE RANGER TRAINING PROGRAM

10. General

This section provides guidance to the officer in charge of the Ranger training program. The training schedules outlined in appendix II contain various combat problems which are presented as a part of the training program. Lesson outlines for these problems are contained in appendix III. Generally speaking, those listed first on the schedule are covered in more detail; those listed last were kept brief to avoid unnecessary duplication. Regardless of those problems selected for the training program, the officer in charge or principal instructor should receive all outlines contained in appendix III.

11. Troop Orientation

a. Troop orientation is given by the principal instructor to familiarize the Soldiers with the general tactical situation, to instruct them in the purpose, and to orient them concerning pertinent safety and administrative details. During this orientation, Soldiers are initially subjected to the realism of the training. From this point on, realism is essential in keeping with the general concepts of Ranger training. It is necessary that the orientation be presented in a realistic manner in order to inject enthusiasm and to stimulate mental attitudes.

b. The instructor presenting this orientation should be designated as a higher headquarters S2 or S3, depending upon the type of mission outlined in the exercise. He should have visual aids present which are normally found in combat under similar circumstances. A general and special situation as an operation order is presented during this orientation. The operations order may be given in its correct sequence or the situation may be presented in a series of notes. The latter method is effective in training an individual leader in arranging his notes into sequence for presentation as his patrol order.

c. After the general and special situation, the patrol members may be required to prepare a warning order and patrol order in writing. The lane instructor then selects an individual who will actually issue the order. He may collect all of the written orders and check them for accuracy. This technique is beneficial in teaching the proper method of preparing a patrol order and does not restrict the training to the patrol leader.

d. The exercises should be tied in with each other in a logical, progressive, overall situation. The continuity of the situation can be maintained by slight alterations in the wording of the general situations presented in each particular exercise.

12. Control Personnel Orientation

This orientation is conducted by the principal instructor to acquaint the control personnel with their duties and conduct, to orient them concerning the play of the exercise, and to explain all pertinent safety and administrative details. It normally takes place at least one day prior to the participating unit's orientation. The control personnel consists of the following:

a. Aggressor control officer.

b. Assistant principal instructor.
c. Supply personnel.
d. Transportation personnel.
e. Aidmen.
f. Aggressor troops.
g. Friendly troops (does not include members of patrol).
h. Lane instructor.

13. Evacuation Plan

Prior to the actual conduct of the exercise, the officer in charge orients all personnel in the location of the medical men, evacuation vehicles, and routes of evacuation. Aidmen are available; however, they do not participate tactically in the exercise. If attached to aggressor units, medical personnel will wear aggressor uniforms. They do not disturb the tactical play of the exercise except for emergencies. Actual injuries incurred are treated under the supervision of the patrol leader. Except in the most extreme cases, care is taken to maintain realism and control personnel should avoid any administrative breaks or halts in the play of the exercise in order to treat the injured.

14. Terrain Preparation

Each problem contains the suggested size of the area and special terrain features required for the exercise. A map reconnaissance of the available terrain establishes areas that appear suitable. A ground reconnaissance determines the best area for the exercise.

a. Friendly Areas (fig. 2). Having selected an area, the location of friendly installations and positions are determined. These positions should be as realistic as possible and in keeping with sound tactical doctrine. Each problem may contain any or all of the following friendly positions, areas, or installations:

(1) Reserve area.
(2) Detrucking point in rear of forward company.
(3) Forward company command post.
(4) Company forward battle area position.
(5) Company outposts.

(*a*) *Reserve area.* The reserve area is located several miles behind the forward positions near a road net. It is concealed from aggressor positions. Tentage, bunkers, foxholes, and various other types of material and equipment may be utilized in the construction of the area so that it affords the appearance of a forward reserve area.

(*b*) *Detrucking point.* This point is concealed from enemy observation and allows a hardstand for vehicles and a suitable area for turn around. There is a prepared po-

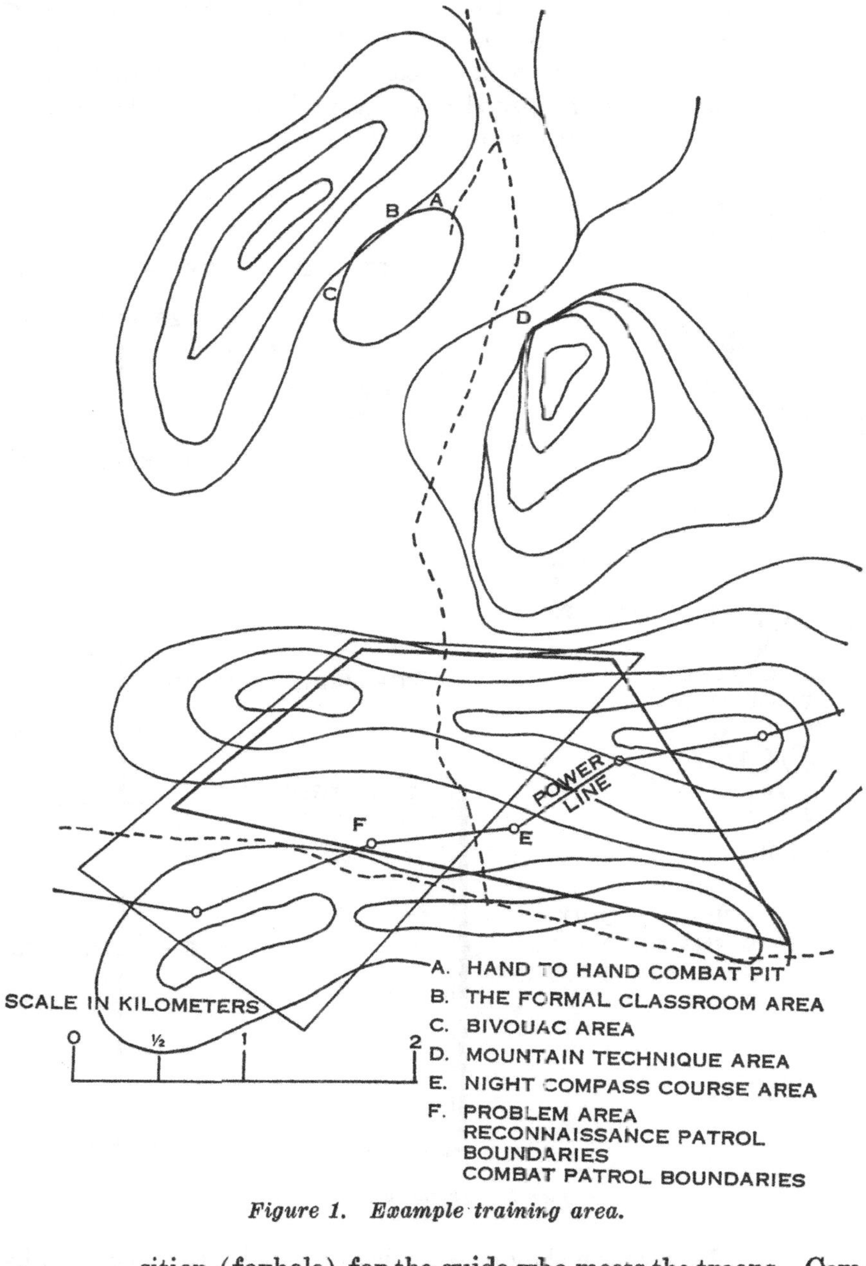

Figure 1. Example training area.

sition (foxhole) for the guide who meets the troops. Company unit signs are conspicuously placed in or near this point. Concealed personnel with simulator artillery bursts or demolition pits are placed near this point to simulate incoming artillery or mortar fire.

(*c*) *Forward company command post.* This position can either consist of a bunker emplacement or foxhole. There should be other equipment such as radios, communication wire, and ammunition boxes placed conspicuously in or near the position to suggest that it is a command post. Again, demolition pits or concealed personnel with artillery simulators are placed near the position to simulate enemy incoming artillery or mortar fire.

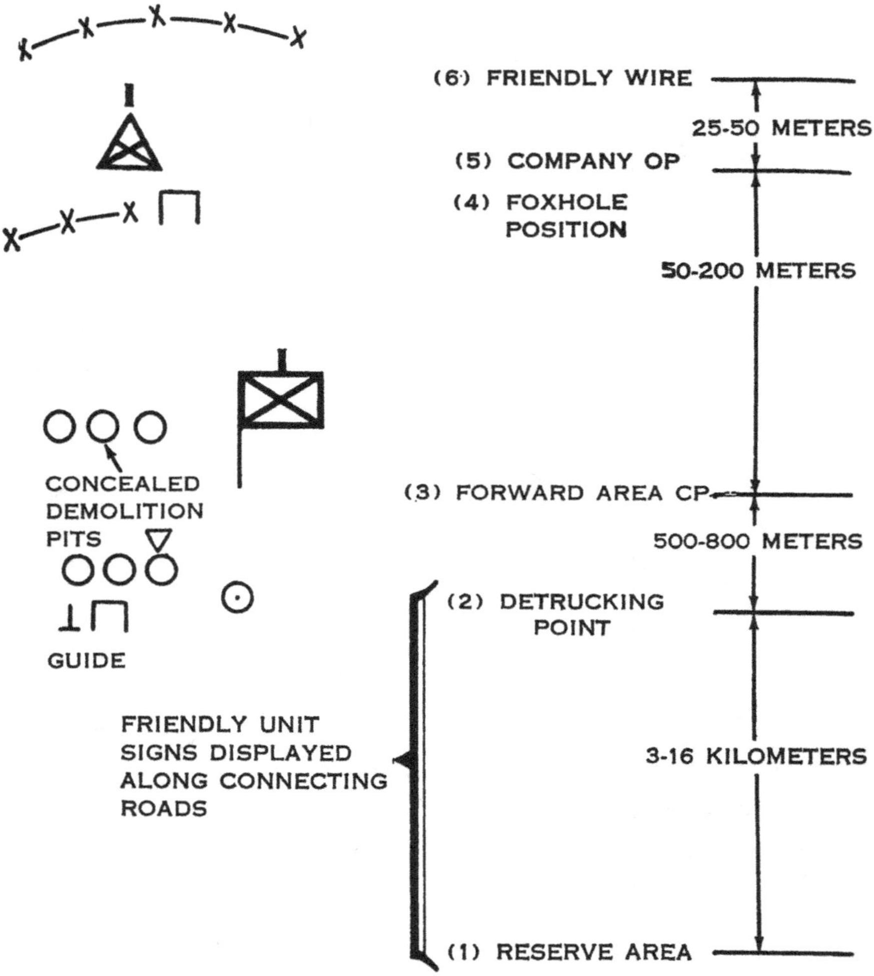

Figure 2. Friendly positions.

The position should be tactically sound. It is occupied by the forward company commander.

(*d*) *Company forward battle area position.* This consists of open foxholes or bunkers, barbed wire or any other materials normally found in and around such a position. All positions are not necessarily manned.

(*e*) *Company outpost.* This position, a foxhole or bunker, is manned by one or two men to give the appearance of a listening post or outpost. Barbed wire is placed in front of the position as would normally be done and a gap exists through which the patrol can depart.

b. Aggressor Areas (fig. 3). The aggressor positions may consist of any or all of the following:

(1) *Outpost.* The aggressor outpost can be constructed in the same manner as the friendly outpost. Roving patrols in

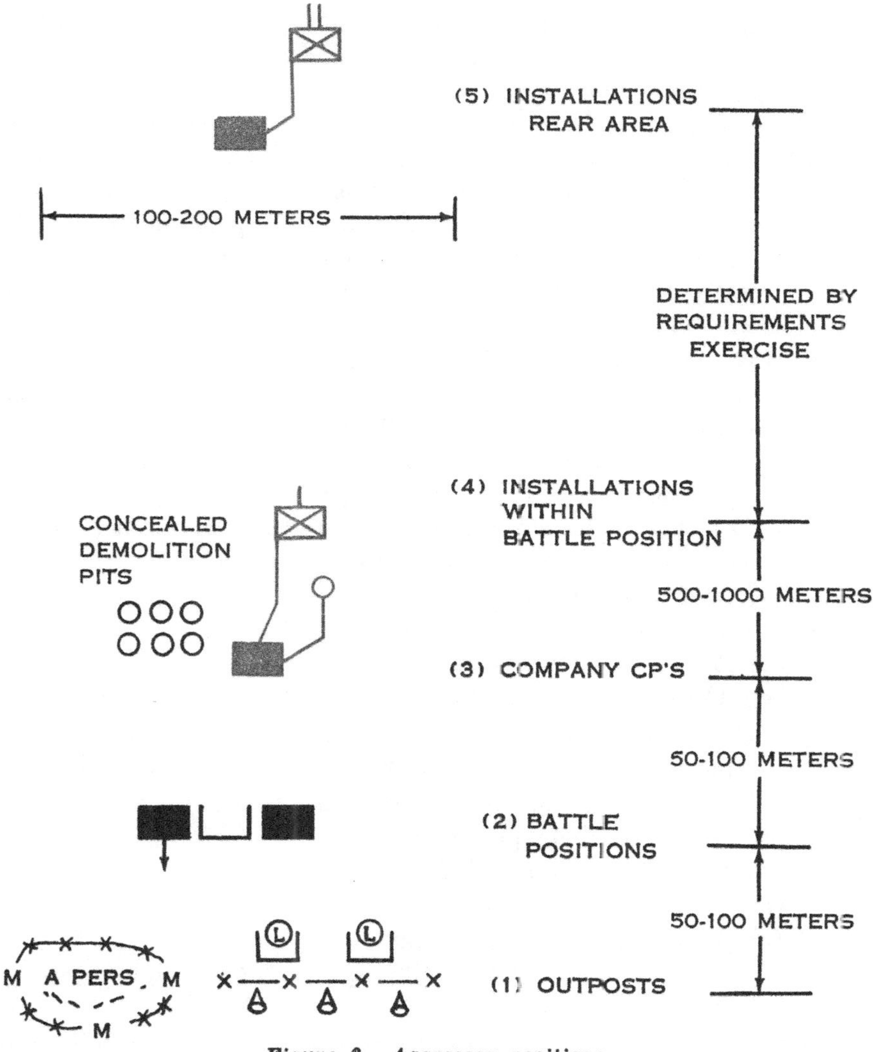

Figure 3. Aggressor positions.

front of the aggressor positions may also be employed in likely patrol avenues not covered by actual emplacements. Dummy marked minefields and barbed wire can be improvised and used to canalize the patrols into occupied positions.

(2) *Main battle position.* These positions, like all other positions and installations, should be located in areas which are tactically sound. Basically, they should be built and occupied to allow for flexibility of movement of available aggressor troops. These troops should be able to move about the position, giving the appearance of a well-manned battle position without making it obvious that there is only a representative group being employed. There should be a lateral road net behind the area selected for the battle position. This allows for rapid movement of troops and facilitates control by the aggressor control officer. Barbed wire should be laid and simulator boobytraps, trip flares, and noisemakers employed.

(3) *Forward company command posts.* These positions can serve as control and supply points for the aggressor. Concealed demolition pits should be near the position to simulate friendly artillery fire. Entrance into the position should be easily gained from the rear to facilitate movement of supplies and personnel.

(4) *Installation within the main battle position.* A variety of installations may be constructed within the position to serve as objective sites. In addition to these installations, there should be aggressor unit signs placed conspicuously along roads and trails to indicate aggressor activity. Deserted buildings and installations may be available for use within the training areas. Examples of these are—

(*a*) Mortar positions.
(*b*) Forward regimental command posts.
(*c*) Artillery installations.
(*d*) Control and checkpoints (traffic and/or security).

(5) *Rear area installations.* These can be treated in much the same manner as those in *a* above. Examples of these are—

(*a*) Regimental and division command posts.
(*b*) Heavy artillery positions.
(*c*) Guided missile positions.
(*d*) Radar installations.
(*e*) Motor pools or supply depots.
(*f*) Critical installations including dams, bridges, and powerplants.

c. Other Areas. This refers to the area between the friendly and aggressor battle positions. The terrain here should remain, as much as possible, in its natural state, thus providing natural obstacles and barriers to the patrol. If possible, there is no activity in this area other than that called for specifically in a particular exercise. This area represents no-man's land and is void of any activity or installations that detract from the realism of the training.

15. Friendly Representation

a. Number. As many of the prepared positions in the forward company area as possible are occupied by friendly troops. In addition to personnel called for specifically in a particular problem the following minimum will, in most cases, be necessary for the occupation of forward positions of one lane (one patrol):

(1) One guide.

(2) One outpost man.

(3) One forward company commander.

b. Uniform. Troops are dressed in the proper United States Army uniform and should give the appearance of having been in combat for some time.

c. Equipment. The individual equipment of the friendly force includes that normally found on a person in a similar capacity during actual combat.

d. Duties and Conduct. The guide's duties consist of meeting the patrol at the detrucking point and leading it to the forward command post and to the outpost.

(1) The forward company commander is prepared to discuss the tactical situation concerning his immediate front with the patrol leader. He is also able to answer all questions and offer aid concerning coordination with the patrol. How much information he volunteers and the amount of coordination he offers should be decided upon by the principal instructor prior to the actual operation of the exercise. See appendix XV, Example Command Post Instructions.

(2) The outpost is generally aware of the tactical situation. A sentry is prepared to coordinate with the patrol leader concerning the time and method of the patrol's return. He is acquainted with the method of challenging and the proper use of the password. See appendix XVI, Example Outpost Instructions.

16. Enemy Representation

a. Number. The number of aggressors is determined by the requirements of the problem.

b. Uniform. Forward of the friendly positions, all personnel, including the principal instructor, aggressor control officer, inspectors, and administrative personnel, are dressed in complete aggressor uniform. It is imperative that this be strictly adhered to and controlled in order to maintain realism and prevent destruction of the tactical atmosphere of the exercise. Care is taken also to have all vehicles forward of friendly lines properly marked with aggressor markings. In keeping with realism, all of the above personnel upon their return into friendly lines revert to the proper United States Army uniform, doing so in such a manner as not to make it obvious that they represent aggressor forces. The same applies to aggressor marked vehicles.

c. Equipment. The aggressor's individual equipment includes those items normally found on a person in a similar capacity during combat.

d. Duties and Conduct.

(1) The employment of aggressor troops depends upon the type exercise being presented. The principal instructor is responsible for the aggressor plan of action. The aggressor control officer is responsible for the proper execution of this plan. This plan includes aggressor action in the main battle position, aggressor rear area, at patrol objective sites, aggressor ambush sites forward of their main battle position, and at any other location where the aggressor is being employed. This plan is presented to the aggressor troops during their orientation.

(2) The action of the aggressor must be logical and, as far as possible, tactically sound in nature. Aggressor tactics are used. In many cases, the enthusiasm of the Soldier depends on the way the aggressor is tactically employed, his enthusiasm, and the way he plays the game.

17. Lane Instructor (LI)

a. Number. At least one lane instructor accompanies each patrol during the conduct of a particular exercise.

b. Uniform. The LI uniform is the same as that designated by the patrol leader in his patrol order.

c. Equipment. Besides the equipment specifically called for in a particular exercise, the LI should have a map of the training area and a compass in order to know at all times the location of the patrol.

d. Duties and Conduct.

(1) During the actual conduct of an exercise, the LI becomes a member of the patrol. He is prepared to give an accurate and critical report on the conduct of the entire patrol at a critique following the exercise. The critique provides the LI with the opportunity to teach. During the actual patrol he offers no criticism or aid to the patrol leader, but he makes mental and written notes of the actions to prepare himself for the conduct of the critique. Only in the case of an emergency, such as the possibility of loss of life, serious injury, or to preclude damage to Government property does he interfere with the patrol leader's decisions or actions.

(2) If desired by the principal instructor, the LI can, throughout the exercise, declare casualties and change the command of the patrol. In doing so, he takes care to maintain the realism and continuity of the exercise.

(*a*) In a case where there is no contact with the aggressor, but the LI desires to change patrol leaders, he might have the patrol leader pretend to break a leg. The second in command or some other person designated by the LI then takes over the command and arranges for this man's disposition. After the new patrol leader's action concerning this man is completed, the LI instructs the casualty to join the other members of the patrol for the continuation of the exercise. The man designated as a casualty simulates being one only so long as the new patrol leader is deciding on and supervising the man's disposition.

(*b*) When the patrol is in contact with the aggressor, the LI may declare casualties as he desires. This is the most advantageous time in which to change command because it tests the new patrol leader's initiative, decisiveness, and other leadership qualities under trying conditions. When the patrol and aggressor actions are stalemated, the LI assesses casualties on both sides in order to force positive action and insure continuation of the exercise.

18. Principal Instructor (PI)

a. Number. There is one principal instructor for each problem.

b. Uniform. The principal instructor's uniform depends upon his location. If he is within friendly lines, it is the proper United States Army uniform. If forward of friendly lines, he wears complete aggressor uniform.

c. Equipment. No special individual equipment is required.

d. Duties and Conduct. The PI is responsible for the orientation of all personnel involved in the exercise and for the preparation of the terrain. He instructs and supervises the aggressor control officer, the LI, the friendly and aggressor representative groups, and all control and administrative personnel. He is responsible for requesting and procuring the necessary communication equipment, ammunition, transportation, rations, and any other equipment necessary to support the exercise. He is also responsible for the proper maintenance of this equipment during the exercises and the proper turn-in of the equipment at the completion of the exercise. He provides for emergency evacuation and instructs all personnel in the operation of this plan. He is responsible for instructing all personnel in pertinent safety regulations and in requiring their adherence to these regulations. The PI supervises the debriefing and critique of the participating troops. Normally, as instructions to various supporting personnel are so lengthy, the PI prepares written instructions. See appendix XIV, Example Lane Instructor Instructions.

19. Aggressor Control Officer

a. Number. One officer or senior noncommissioned officer is designated aggressor control officer.

b. Uniform. While forward of the friendly positions the aggressor control officer wears proper aggressor uniform.

c. Equipment. Although it is not absolutely necessary, it is desirous that the aggressor control officer be armed and have a map of the training area.

d. Duties and Conduct. The aggressor control officer is responsible for the proper execution of aggressor actions as outlined by the principal instructor in the orientation. He assists the principal instructor by supervising the aggressor troops (including administrative personnel within the positions) in preparing their position, maintaining their equipment, adhering to safety regulations, execution of the evacuation plan if necessary, and the aggressor play of the problem.

20. Safety Personnel (Aidmen, Roadguards, Ambulance Driver)

a. Number. The number and types of safety personnel are designated for each individual exercise.

b. Uniform. Uniform, depending on the men's location, is either United States Army or aggressor.

c. Equipment. Safety personnel will have the special equipment necessary to accomplish their assigned task. In the case of an aidman, an aid bag and litters may be desired. Ambulances or air vehicles within aggressor positions are marked as aggressor vehicles.

d. Duties and Conduct. Roadguards are placed tactically in the play of the problem regardless of whether they are within friendly or aggressor positions. Safety personnel maintain the realistic continuity of the problem as far as possible.

21. Supply Personnel

a. Number. One noncommissioned officer is designated as supply noncommissioned officer for the exercise.

b. Uniform. United States Army uniform.

c. Equipment. No special equipment is required.

d. Duties. The supply noncommissioned officer assists the principal instructor in requesting, procuring, issuing, and receiving all supplies necessary to support the exercise. He insures that the equipment is serviceable prior to issue and that it is properly cleaned and serviced prior to turn-in. He reports all items lost or damaged. He insures that safety regulations are observed when issuing, receiving, and storing explosives and ammunition. The supply noncommissioned officer and assistants also assist in maintaining the problem realism by providing sound logical reasons to the Soldier when equipment requested is not available.

22. Transportation

a. Number. The vehicular requirement is based on the number of vehicles necessary to support the problem. Availability of vehicles is a consideration.

b. Uniform. Uniform, depending on the driver assignment, is either United States Army or aggressor.

c. Equipment. No special equipment is required.

d. Duties and Conduct. Vehicle drivers do not participate tactically in the exercise with the exception of being dressed in the proper uniform and having the vehicles marked properly.

23. Communications

Radios, wire, or other methods of communication are integrated into all the exercises for control and safety purposes. A control net consisting of friendly personnel and aggressor personnel, each with different call signs, can be utilized. A separate net may be established for the patrol. In the case of the patrol net, the principal instructor establishes a station representing friendly elements to answer the patrol's calls and to communicate with them. See appendix X.

24. Critique

a. Purpose. The critique provides an opportunity to correct errors. It also provides an opportunity to compare exercises.

b. Conduct. Make the critique constructive.

(1) Briefly reviewing the action.
(2) Point out the Soldiers' achievements during the exercise.
(3) Point out the major errors noted and give suggestions.
(4) Encourage the Soldier to ask questions that will clarify his understanding.
(5) Summarize the lessons learned.
(6) Create in the Soldiers a feeling of accomplishment and a desire for continued achievement in training.

c. Conferences. Generally, the critique can be given most effectively by conference because this method encourages a two-way exchange of ideas and thoughts between the lane instructor and the Soldiers. Guard against antagonizing and discouraging the group. Do not present a long list of deficiencies; avoid strong criticism of an individual or a unit in the presence of the entire group. Sometimes it is advantageous to conduct several critiques—one for the unit and a separate critique for the patrol leaders. This avoids possible resentment or lowered morale of any Soldier or units taking part in the exercise.

d. Offer Solutions. Keep in mind the purpose of the training exercise. Recognize those who make outstanding contributions to their team's performance, and call attention to any errors or incorrect tactics without becoming personal. When errors are noted, give the correct solutions. When more than one solution is possible, give a preferred solution. Emphasize that other solutions are permissible, provided the fundamental points are correct and sound principles are followed.

e. Checklists. The officer in charge should encourage PI's to prepare checklists for use by the lane instructors. An example checklist is contained in appendix XI.

25. Grading

a. When an evaluation system is utilized in conjunction with the training program (par. 28), the principal instructor establishes the grading phases for the problem. The lane instructor rates the performance of the various patrol leaders. The lane instructor can do this, along with his other duties, regardless of the number of leaders designated during the conduct of one of the exercises. The leader is graded on that phase in which he served in a command capacity. If one individual remains the leader throughout the entire exercise, only one grade is necessary. Two patrol leaders during the exercise of one patrol constitute the basis for two grades, etc. If more than one patrol is run during a single field exercise, several grades from each patrol may be obtained.

b. If the principal instructor desires to control the number of grades and the time an individual is to remain in a leadership capacity, he may assign a specific number of grades to be obtained. He does this by designating "phase lines"; points where leaders can be logically changed. The lane instructor can then be instructed to change the leaders at a given location (phase line) prior to the actual conduct of the exercise (fig. 4). To obtain four patrol leader grades during one patrol, the "phase lines" are assigned as follows:

(1) Patrol leader No. 1 (phase I). Leads patrol initially and is relieved upon the patrol's arrival at the friendly forward positions.

(2) Patrol leader No. 2 (phase II). Leads patrol through the friendly forward areas and to the objective.

(3) Patrol leader No. 3 (phase III). Conducts the action at the objective.

(4) Patrol leader No. 4 (phase IV). Conducts withdrawal from objective and leads patrol back to friendly positions.

c. When more than one patrol is being run in the same exercise, the lane instructor with each patrol is required to get four grades. Figure 5 shows a method of obtaining six grades. In designating dif-

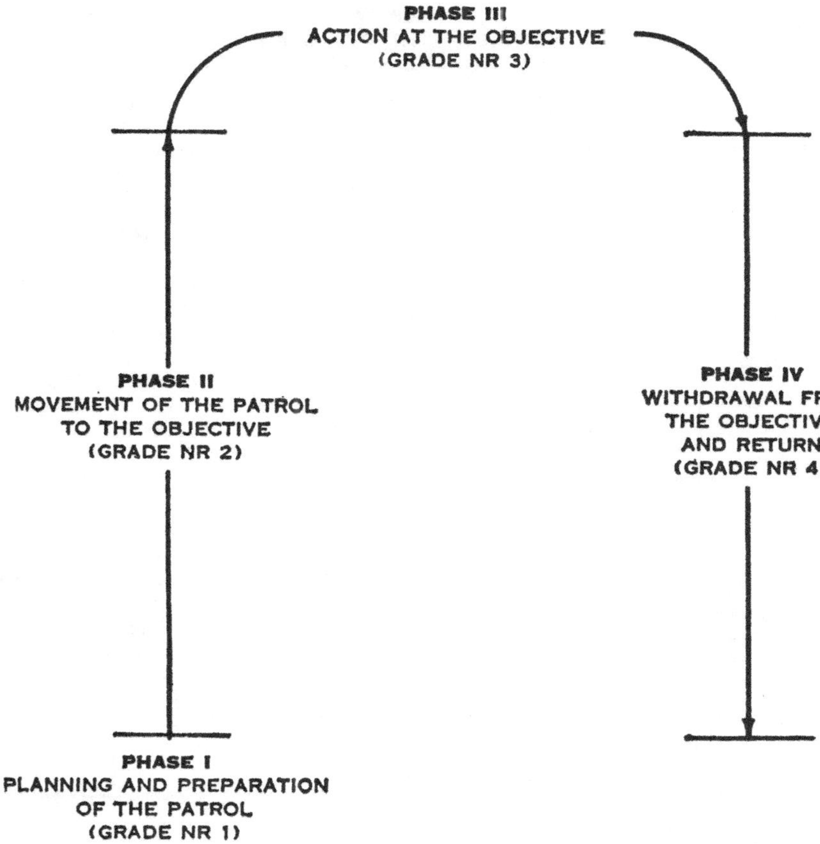

Figure 4. Phase I including grading for one patrol (four grades).

ferent patrol leaders by use of the "phase line" system, realism should be maintained. When a patrol leader is relieved, he should be declared a casualty by simulating explosion of a mine or artillery.

26. Equipment Checklists

Various patrol missions require different types and quantities of equipment. The following lists suggest normally obtainable items for different patrol missions. Patrol leaders must determine the quantity suitable to the mission, terrain, and requirement.

a. Reconnaissance Patrol.

Binoculars.
Camouflage sticks.
Compass.
Wire cutters.
Flashlights.
Pocket or hunting knives.
Machetes.
Metascopes.

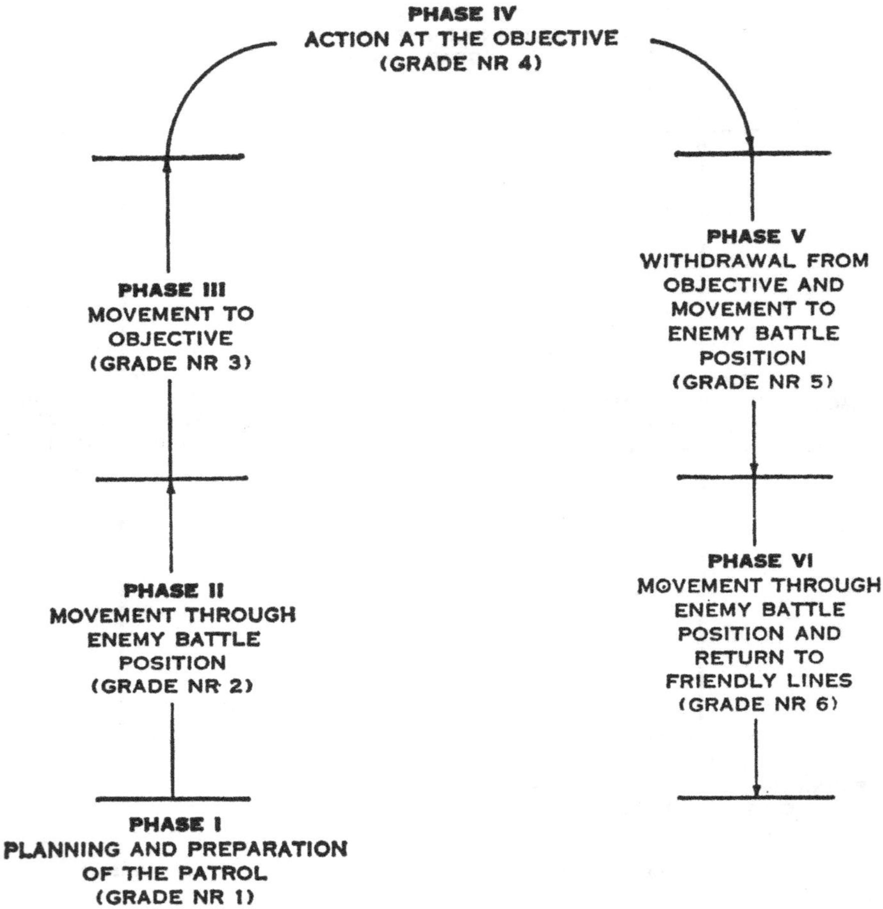

Figure 5. Phase line grading for one patrol (six grades).

Radio (AN/PRC–6 or AN/PRC–10).
Rope, safety or climbing, 7/16-inch nylon.
Tape, luminous.
Whistle, thunderer.
Snakebite kit.
Grenades (fragmentation, white phosphorous).
Sniperscope.
Flares and projector.
Pencil and paper.
Oil, thong, gun patches.
Automatic rifles.
First aid kit.
Pneumatic raft.
Maps.

b. Combat Patrol. Consider the equipment list for a reconnaissance patrol and add the following items for consideration:

Blasting caps.

Crimpers.
Demolitions.
Entrenching tool.
Fuze cord.
Machineguns.
Launcher, rocket, 3.5-inch.
Mortars.
Grenade launchers.
Packboard.
Friction tape.
Steel helmet w/liner.

c. Ambush Patrol. Consider reconnaissance patrol and combat patrol equipment lists and add the following items for consideration:

Mines, AT, AP, Claymore.
Recoilless rifle (56- or 75-mm).
Grenades (Thermite, Energa).
Telephones, TA–1/PT.
Wire, WD–1.
Wire, barb, and pickets.

d. Summary. The above equipment is in addition to individual equipment and weapons normally carried. It must be emphasized that this is not a complete list. As new items of issue become available, they should be added to the list. Patrols need not carry all the items indicated. Miscellaneous items which are compact, convenient, and present no carrying problem and which may prove invaluable in the event of evasion or survival techniques should also be considered. These and other items are listed in paragraph 53.

27. Evaluation

a. The officer in charge of the training program should consider using some sort of evaluation system. Some procedure for observing and reporting the individual's progress is necessary. Under the unit training system, utilizing leaders' verbal reports might suffice. Normally, Ranger units or other organizations training for special missions will not have the time to effect an evaluation system. If the subordinate leader states that a Soldier is not suited for certain operations, then he must be eliminated on the strength of this report.

b. Under an individual training program, especially in peacetime, the officer in charge will need written reports to support his recommended Soldier disposition, evaluate his instruction, and establish specific training standards.

c. For the most part, grading in a Ranger training program is subjective. To attain maximum standardization and objectivity, all concerned must—

(1) Know the objective of the overall evaluation system.

(2) Know the approved or accepted solution to a problem, but base Ranger Soldier evaluation on the feasibility of the Soldier's solution.

(3) Use all possible checklists (problem checklists, warning order, patrol order, estimate of the situation, and troop leading cards) to assist in grading.

(4) Look for individual attributes as well as deficiencies and weigh one against the other.

(5) Submit observation reports that state accurately what the Soldier did or failed to do; be factual, objective, and avoid inconsequential matters. Use personal opinion only in the summary paragraph.

d. To attain a true evaluation of each Soldier's performance and Ranger potential the following qualities should be considered:

(1) *Confidence.* Firmly believing in one's own ability to surmount obstacles successfully. Ability to overcome inherent fears with minimum hesitation.

(2) *Discipline.* Striving to do a good job as a leader or follower. Prompt and cheerful execution of both the letter and spirit of orders regardless of personal environment. The ability to accept counselling without rancor (self-discipline). The ability to place the mission before self-comfort. Instinctively doing what is right under all conditions.

(3) *Endurance.* The drive, fortitude, self-discipline or demonstrated physical stamina to push oneself to the limit, to carry on mentally and physically regardless of fatigue, hunger, and the pressures of a combat situation.

(4) *Technical knowledge.* The proper and effective use of weapons and equipment in accomplishing the assigned mission.

(5) *Tactical knowledge.* Application of appropriate principles, techniques, and commonsense in reaching a decision or initiating courses of action to include the timeliness of this action.

(6) *Leadership.* The application of principles of leadership and troop leading procedures. The ability to develop team spirit, accomplish the mission, set the example, function as a teamworker, demonstrate good Soldierly habits, initiative, enthusiasm, attention to detail, and dependability.

28. Evaluation System

a. Points. A 1,000 point block is usually the easiest to administer. This system divides these points into six subareas, with point weights proportionally allocated to overall importance of the area.

Patrol grades	500
Spot reports	100
Tactical officer's rating	150
Map reading and compass course	100
Ranger evaluation report	100
Other tests	50
Total	1.000

b. Patrol Grades (figs. 6 and 7).

(1) These grades are given to Soldiers placed in a common position on graded patrol problems and are the most weighted

RANGER Smith (Last Name) Arthur (First Name) C. (MI) DATE 26 January ——

PROBLEM 8244 CAPACITY Patrol Leader PHASE

1. OBSERVATIONS:

A. Warning Order

1. Manner of delivery -- Fair

Lack of Force

Read order directly from paper - never glanced up - used notes as a crutch rather than aid.

Due to the above lost interest of group - men sleeping and talking.

2. Format

Enemy situation too detailed. Did not talk about specific area of operations.

Equipment - Fair

Insufficient maps, only one machette, no radio requested

Specific instructions for subordinates

Failed to give second in command detailed instructions for preparation of the patrol.

Time Schedule - good except time of evening meal was omitted.

B. Patrol Order

1. Manner of delivery - Good

But still lacked force

2. Format - Good

Enemy and friendly situations not detailed enough.

Execution - Good

Alternate routes of return not covered.

Actions at rally points - could have been more detailed.

Command and Signal - O. K. but failed to give intra-patrol password.

EXCELLENT	90-100	
SATISFACTORY	70-89	70
UNSAT	0-69	

LT JOHN MOORE
PRINT NAME & GRADE

Date Counselled

Figure 6. Example patrol grade (front).

623989°—62——3

C. Rehearsal - Unsatisfactory
Area selected - poor, no attempt to indicate objective
Control - weak - men were joking around - not paying attention
Did not have evening rehearsal
SUMMARY:
Ranger Smith displayed a fair knowledge of patrol techniques but lacked the force necessary to get things done. Needs more work in controlling and communicating with men. Very unimpressive, hesitates to impose his will upon others.
INSTRUCTIONS FOR USE
a. Record factual information about what the student did or failed to do.
b. Summarize the overall student performance.
c. Enter the Numerical grade awarded.
d. The information stated in this report will be used as a basis for counselling the student should he receive an unsatisfactory grade. Lane observers will include what the student did or failed to do during his critique.
e. Report is signed by the lane observer and submitted to the tactical officer.
f. The tactical officer will sign and date counselling block if and when student is counselled.
SUMMARY

Figure 7. Example patrol grade (back).

in the evaluation system. Every effort is made to insure that they are accurately awarded and substantiated by factual reports.

(2) To compute, add numerical scores awarded on all patrol observation reports, divide the total score by the number of grades, multiply the quotient by five.

c. Spot Reports (fig. 8).

(1) These reports are established to provide an additional means of evaluating performance. They are rendered beginning with the third day of training. Spot reports are primarily intended for evaluation of practical work performance when in a secondary position of responsibility; however, they may

be used for repeated misconduct or deportment. It is not intended that they be used to report routine "gigs" which should be reflected in the tactical officer's rating. Care should be taken not to negate the value of the report by indiscriminate use.

RANGER	Smith (Last Name)	Arthur (First Name)	C. (MI)	DATE	26 August ——
PROBLEM	#6	CAPACITY	Patrol Member	PHASE	Objective Area

1. OBSERVATIONS:

Ranger Smith was observed by the undersigned arguing in a loud voice with the patrol leader at the objective area. This excessive noise alerted the guerrilla camp as to the presence of the patrol.

INSTRUCTIONS FOR USE

a. Write in word "Spot" in large print at bottom of page.

b. Enter student's name, date, problem number, job capacity and phase if applicable.

c. Place an X opposite adjective rating award.

d. In a few words state why rating was awarded.

e. Sign and submit to tactical officer.

f. Tactical officer will counsel student if he receives an excessive number of unsatisfactory spot reports.

~~EXCELLENT~~	XXXXXXX	
~~SATISFACTORY~~	XXXXXXX	
UNSAT X	XXXXXXX	-10

LT JOE BROOMFIELD
PRINT NAME & GRADE

Date Counselled ____________

SPOT

Figure 8. Example spot report.

(2) Only excellent and unsatisfactory spot reports become part of the record. Each Soldier is given 100 points which he must protect or strive to increase.

(3) Spot reports are computed on a plus or minus basis and carry

the following weights to be added or subtracted to the total points allocated:

Excellent	+10 points
Unsatisfactory	—10 points
Initial allocation	100 points
Maximum score	No limit
Minimum score	0 points

d. Tactical Officer's rating (fig. 8).

(1) At the end of specific periods, the tactical officer rates each Soldier in the class. Maximum available time must be spent in getting to know each man to insure that the rating reflects firsthand knowledge of the individual's performance and potential. The same form used for the patrol and spot reports is used to record the tactical officer's grade and remarks.

(2) To compute, average the total tactical officer's grades and multiply by 1.5.

e. Map Reading and Compass Course Grade.

(1) Each Soldier is required to take a written map test and to negotiate a compass course during the first week of training. Those who score less than 50 percent on the compass course must take a retest. The scores attained on the map test and the first running of the compass course are used in computing class standings. A failure on the second compass course test may be made a basis for elimination from the training.

(2) To compute, map and compass scores are averaged.

f. Ranger Evaluation Report (fig. 9).

(1) At the end of certain training phases, every Soldier rates every other man in his squad. The individual is required to write a brief word picture on fellow squad members and rate them as to standing within a forty man group.

(2) To compute, evaluation reports are averaged.

g. Other Tests. The contents of these tests will vary with the subjects taught. They are useful to determine the effectiveness of the instruction and assist in the evaluation of the Soldier.

h. Miscellaneous Records. Generally all written information on the Soldier should be contained in his record. Some items to consider in addition to those above are—

(1) Physical fitness test scores.

(2) Sick slips (number of training hours missed).

(3) Any counselling Soldier has received. A counselling form may be used.

(4) Background on Soldier, i.e., length of service, etc. A Soldier's personal history form may be appropriate.

(5) Board reports, if this procedure is used, for Soldier elimination.

RANGER EVALUATION REPORT

INSTRUCTIONS:

1. Evaluate each man in your squad in comparison within a 40-man Ranger group. Consider the characteristics listed below prior to determining the man's rating. Every student assigned to your squad must be rated.

2. Under remarks state briefly characteristics (desirable or undesirable) of this man that impressed you most.

EVALUATED STUDENT'S NAME (Last, First)	Date DAY / MONTH / YEAR

RANGER CHARACTERISTICS

TACTICAL KNOWLEDGE	GOOD SOLDIERLY HABITS	CONFIDENCE
PHYSICAL STAMINA	SELF DISCIPLINE	ENTHUSIASM
TIMELINESS OF ACTION	DRIVE	INITIATIVE
ATTENTION TO DETAIL	TEAMWORK	DEPENDABILITY

PHASE OF TRAINING (Circle One)	BENNING	FLORIDA	MOUNTAINS	STANDING WITHIN A 40-MAN GROUP (Circle One)

REMARKS:

1.	11.	21.	31.
2.	12.	22.	32.
3.	13.	23.	33.
4.	14.	24.	34.
5.	15.	25.	35.
6.	16.	26.	36.
7.	17.	27.	37.
8.	18.	28.	38.
9.	19.	29.	39.
10.	20.	30.	40.

SIGNATURE	GRADE	SERIAL NO.

Figure 9. Example evaluation report forms.

29. Weapons, Equipment, and Ammunition

a. Weapons and Equipment.

(1) The equipment, supplies, and materials available in most Table of Organization and Equipment (TOE) organizations are normally sufficient to conduct a Ranger training program: however, it may be necessary to requisition some equipment above TOE or Table of Allowance (TA) authorization. Authority for this equipment will vary between commands. The headquarters initiating the training should outline requesting procedures to include authority to the officer in charge.

(2) Equipment required to conduct Ranger type training is classified in three groups, TOE, TA, or Special and Pre-

fabricated. Requisitions for items not on hand should be projected at least 60 days prior to the start of the program.

(3) In *b* through *e* below are lists showing an approximate recapitulation of required training equipment and material to support the training outlined in this manual. Total figures are based on a Soldier group of 100. Variations in the problems to fit the terrain and objectives may change the number and types of equipment required; however, the instructor will find the following paragraphs a convenient checklist.

b. TOE Equipment.

Item	Amount		
	1-week	2-week	3-week
ENGINEER:			
Blasting machine	1	1	1
Compass lensatic 5° damped	110	110	110
Crimper (cap demo)	2	25	25
Galvanometer	1	1	1
Metascope (optional)	(*)	(*)	(*)
Shovel, D-handle	10	10	10
Sniperscope	(*)	(*)	(*)
ORDNANCE:			
Binocular, M7	12	12	12
Automatic rifle	10	10	10
Grenade launcher	35	35	35
Machinegun	15	15	15
Launcher, rocket 3.5-inch (optional)	0	(*)	(*)
SIGNAL:			
Flashlight MX-991	30	30	30
Radio AN/PRC-6	15	15	15
Radio AN/PRC-10	25	25	25
Radio, vehicle mounted (10–15 mile radius)	3	3	3
Sound equipment set AN/TIQ-2	1	1	1
Telephone TA 312/PT	(*)	(*)	(*)
Wire, WD-1/TT (Mine)	(*)	(*)	(*)
TRANSPORTATION:			
Ambulance ½- or ¾-ton	2	2	2
Ammunition vehicle	1	1	1
Truck ¼-ton	5	5	5
Truck 2½-ton	7	7	7
Aircraft	0	as req	as req
QUARTERMASTER:			
Axe, single bit 4-lb	3	3	3
Cutter, wire M1938	10	10	10
Lantern, gasoline, leaded fuel	2	2	2
Packboard, plywood w/loading rope	10	10	10
Whistle	6	6	6

c. TA or Special Equipment.

Stock No.	Nomenclature	Amount		
		1-wk	2-wk	3-wk
ENGINEER:				
8105–285–4744	Bag, burlap, sand, drawstring closure; 26½″ long, 14″ wide	1, 500	2, 000	4, 000
9505–251–4482	Barbed wire; steel, galvanized rd 0.155 in. dia, 2 strand; 4 points 0.0800 in. dia, 4″ between center 100 lb spool	15	as req	as req
Nonstandard	Cap (demo dummy)	10	10	10
9505–371–9494	Concertina wire, barbed, roll	as req	as req	as req
4210–223–9909	Extinguisher, fire, carbon tet, hand-trigger controlled	1	1	1
5110–223–6260	Machette, rigid handle 18″ lg 2¼″ w/blade, 5¼″ hdle	2	2	10
8510–161–6204	Paint, face, camouflage, light green, and loam	20	25	30
5120–223–8426	Digger, post hole hinged, two handles 60″ long	1	1	1
5110–263–8836	Saw cross-cut 2-man 5 ft lg	1	1	1
ORDNANCE:				
Nonstandard	Adapter, blank, gun, machine	20	20	20
Nonstandard	Adapter, blank, rifle	200	200	200
Nonstandard	Adapter, blank, automatic rifle	10	10	10
	Mortar, 60-mm (optional)	(*)	(*)	(*)
	Rifle, recoilless 57-mm (optional)	(*)	(*)	(*)
	Rifle, recoilless 75-mm (optional)	(*)	(*)	(*)
SIGNAL:				
Not applicable	Projector, film 16-mm	1	1	1

Stock No.	Nomenclature	Amount		
		1-wk	2-wk	3-wk
QUARTERMASTER:				
8415–268–7875	Gloves, leather, work type, mens gauntlet cuff	105	105	105
5120–255–7560	Hammer, hand, Piton, mountain, 9 oz head weight, MIL–H–1431	3	3	3
5120–224–4130	Hammer, hand, sledge, blacksmith cross peen 12 lb	2	2	2
4020–281–4101	Marline, hemp, tarred, roll 180 ft per lb	0	1	1
8465–240–2971	Piton, mountain, steel, horizontal	50	50	50
8465–240–2972	Piton, mountain, steel, short vertical	50	50	50
8465–240–2973	Piton, mountain, steel, wafer	50	50	50
8465–240–2974	Piton, mountain, steel, tubular	2	2	2
8465–240–2975	Piton, mountain, steel, angle	50	50	50
4020–231–2581	Rope, Manila, ½″ dia 3 strand, preservative oil treated	500	500	500
4020–238–7732	Rope, Manila, 1″ dia, 3 strand, preservative oil treated	500	500	500
4020–242–1069	Rope, Manila, ¼″ dia, 3 strand, preservative oil treated	2, 400 ft	2, 400 ft	2, 400 ft
4020–231–2537	Rope, nylon O.D. 3 strand, 1¼″ cir, 12.9 to 14.3″ per 10 turns 120 ft overall length	16	21	26
8465–360–0228	Snap link, mountain, Piton, steel, zinc plated Type I	115	115	115
Nonstandard	Table (demo inst)	0	10	10
4970–644–3167	Tape, elec, friction, black ¾″ x 82.5 ft	15	25	35
59–8940–400–200	Tape, phosphorescent, 4″ x 20 yd roll	1	1	1
4020–247–1735	Twine, ctn, Z-twist preservative wax treated 4 oz ball	2	2	2
N/A	Tentage (briefing, supply, cadre) officers and EM—opns	(*)	(*)	(*)
MEDICAL:				
	Litter (instructional purposes)	1	1	1

d. Prefabricated and Miscellaneous Equipment.

Item	Amount		
	1-wk	2-wk	3-wk
Aggressor uniforms	100	100	100
Aggressor vehicle signs	(*)	(*)	(*)
Bear claw (cliff assault)	0	0	1
Blackboard	2	2	2
Chart blowup of problem areas	(*)	(*)	(*)
Chalk and erasers	(*)	(*)	(*)
Class and instructor name cards	(*)	(*)	(*)
Compass stakes (box for cards)	25	25	25
Compass aiming stakes	2	2	2
Confidence test	1	1	2
Dummy explosive charges	(*)	(*)	(*)
Grappling hook (cliff assault)	0	0	1
Material for uniform tiedowns	(*)	(*)	(*)
Pulley rope	(*)	(*)	(*)
Pointer (instructional)	2	2	2
Pace course marker	3	3	3
P.T. platform	1	2	2
Rope bridges and traverses	(*)	(*)	(*)
Problem overlays	(*)	(*)	(*)
Podium	2	2	2
Sound table	(*)	(*)	(*)
Scaling ladders	(*)	(*)	(*)
Snakebite kits	(*)	(*)	(*)
Spikes, assorted nails	(*)	(*)	(*)
Sawdust pit	1	1	1
Safety boats	(*)	(*)	(*)
Rappel seat rope	50	50	50
Wing tanks, 106 RR tubes or other items for objective area	(*)	(*)	(*)

Note. Miscellaneous items—some equipment can be fabricated, such as the modified snap link shown in figure 10.

MANY ITEMS MAY BE FABRICATED IN BATTLE GROUP OR DIVISION ORDNANCE SHOPS, SUCH AS THE MODIFIED SNAP LINK PICTURED BELOW:

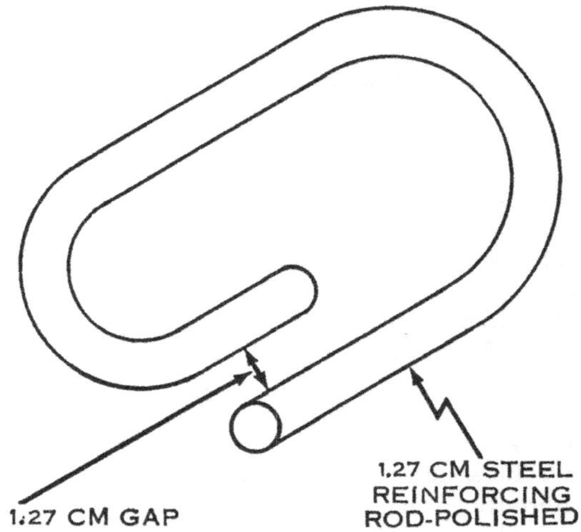

Figure 10. Prefabricated snap link.

e. Ammunition. Authority for ammunition is usually contained in TA 23–100 for most commands. The following is a listing of ammunition to support the training programs outlined in this manual:

Note. One asterisk (*) indicates—total varies. Double asterisk (**) indicates—problem 8 not included.

Types of ammunition problem No.	Number of items per student				Number of items for demonstration and/or practical exercise			
	Breakdown by problem	Totals			Break-down by problem	Totals		
		1-wk	2-wk	3-wk		1-wk	2-wk	3-wk
1	2	3	4	5	6	7	8	9
0280								
Gren Hand Smk HC		1 per 15	1 per 15	1 per 15				
Problem #2	1 per 15							
Problem #3	1 per 15							
Problem #4	2 per 15							
Problem #6					4	0	0	4
1020								
Adaptor Priming Explosive								
Problem #8					5	0	(*)	5
1070								
Block Demolition 2½ lb Plastic				24 per 37				
Problem #3								
Problem #4								
Problem #8	8 per 30	0	(*)		15	0	(*)	15

Types of ammunition problem No.	Number of items per student				Number of items for demonstration and/or practical exercise			
	Breakdown by problem	Totals			Break-down by problem	Totals		
		1-wk	2-wk	3-wk		1-wk	2-wk	3-wk
1	2	3	4	5	6	7	8	9
1150								
Cap Blasting Elec Spec	2 per 25	4 per 25	4 per 25	64 per 28				
Problem #2	2 per 25							
Problem #3	2 per 25							
Problem #4	2 per 25							
Problem #5	15 per 30							
Problem #8	45 per 30				26	0	(*)	26
1170								
Cap Blasting Nonelec Spec				24 per 15				
Problem #4							(**)	
Problem #8	24 per 15	0	(*)		20			29
1200								
Charge Shaped 15 lb								
Problem #8		0	0	0	5	0	(*)	5

1210								
Charge Shaped 40 lb								
Problem #8		0	0	0	1	0	(*)	1
1230								
Clip Cord Detonating								
Problem #8		0	0	0	5	0	5	5
1240								
Cord Detonating Petn	10′ per stu			10′ per 1				
Problem #8		0	(*)		500	0	(*)	600
1305								
Ctg Blank Cal .30	168 per 1	296 per 1	400 per 1					
Problem #1	80 per 1							
Problem #2	80 per 1				720			
Problem #3	80 per 1				400			
Problem #4	128 per 1							
Problem #5	80 per 1							
Problem #6	24 per 1							
Problem #9	8 per 1				48	768	448	448

Types of ammunition problem No.	Number of items per student				Number of items for demonstration and/or practical exercise			
	Breakdown by problem	Totals			Break-down by problem	Totals		
		1-wk	2-wk	3-wk		1-wk	2-wk	3-wk
1	2	3	4	5	6	7	8	9
1310								
Ctg Blank Cal .30 Ctn								
Problem #2					1200			
Problem #3					1200			
Problem #4					1200			
Problem #5					1200			
Problem #6					1200	1200	3200	5600
1320								
Ctg Blank Cal .30 MLB			165 per 1	260 per 1				
Problem #1								
Problem #2	50 per 1							
Problem #3	40 per 1							
Problem #4	125 per 1							
Problem #5	100 per 1							
Problem #6					500			
Problem #9					500	1000	500	1000

1690								
Explosive Cratering 40 lb					10	0	(*)	10
1730								
Explosive Tetryl								
Problem #8	¼ lb per stu		(*)	(*)	3	0	(*)	3
1710								
Explosive TNT ½ lb Block								
Problem #1	2 per 25							
Problem #2	2 per 25							
Problem #3	2 per 25							
Problem #4	2 per 25							
Problem #5	1 per 2							
Problem #8	1 per stu		(*)	(*)	52		(*)	52
1720								
Explosive TNT 1 lb Block								
Problem #8	1½ per stu			1 per stu	47	0	(*)	47
1730								
Firecracker					200			
Problem #2								
Problem #4	4 per stu				6	200	6	6

Types of ammunition problem No.	Number of items per student				Number of items for demonstration and/or practical exercise			
	Breakdown by problem	Totals			Break-down by problem	Totals		
		1-wk	2-wk	3-wk		1-wk	2-wk	3-wk
1	2	3	4	5	6	7	8	9
1770								
Flare Trip				1 per 5				
Problem #4					12			
Problem #5	1 per 10	0	0		10		12	22
1780								
Fuze Blasting Time								
Problem #2	1 per 15	1 per 15	(**)	60 per 15				
Problem #4					8			
Problem #8	60 per 15		(**)		100	0	8	108
1895								
Lighter Fuze Waterproof								
Problem #2	1 per 15	1 per 15	(*)	23 per 15				
Problem #4					8			
Problem #8	23 per 15				15	0	8	23

3320								
Sig Ground Green Star Cluster		2 per 30	2 per 30	2 per 30				
Problem #2	2 per 10							
Problem #3	1 per 30							
Problem #4	1 per 30							
Problem #5	3 per 30							
3350								
Sig Ground Red Star Cluster:								
Problem #2	2 per 30	2 per 30	5 per 30	5 per 30				
Problem #3	3 per 30							
Problem #4	2 per 30							
3370								
Sig Ground White Star Cluster:								
Problem #1	1 per 30	3 per 30	3 per 30	3 per 30				
Problem #2	2 per 30							
Problem #3	1 per 30							
Problem #4	1 per 30							
3380								
Sig Ground White Star Prcht:								
Problem #4					100	0	100	100

Types of ammunition problem No.	Number of items per student				Number of items for demonstration and/or practical exercise			
	Breakdown by problem	Totals			Break-down by problem	Totals		
		1-wk	2-wk	3-wk		1-wk	2-wk	3-wk
1	2	3	4	5	6	7	8	9
3390								
Simulator Boobytrap Flash		1 per 20	3 per 20	4 per 20				
Problem #2	1 per 20							
Problem #3	1 per 20							
Problem #4	1 per 10							
Problem #5	1 per 20							
3400								
Simulator Boobytrap Illum				1 per 6	5			
Problem #1					10			
Problem #2								
Problem #3								
Problem #5	1 per 6	0	0			15	15	15
3420								
Simulator Grenade Hand		21 per 10	31 per 10	51 per 10				
Problem #1	1 per 1							
Problem #2	1 per 1				50			

Problem #3	1 per 1							
Problem #4	1 per 1							
Problem #5	1 per 1							
Problem #6	1 per 1							
Problem #9	1 per 10				24	74	24	24
3440								
Simulator Shell Burst Ground		2 per 10	4 per 10	10 per 10				
Problem #1					20			
Problem #2	2 per 10							
Problem #3	2 per 10							
Problem #4	2 per 10							
Problem #5	6 per 6							
Problem #6					40			
Problem #9					12	32	32	72
3460								
KIT, Torpedo Bangalore								
Problem #8					7	0	(*)	7

30. Training Safety

Troop safety is a vital part of a Ranger training program. To insure realism, safety procedures established as part of a Ranger training program should not interfere with the tactical play of problems. Every effort should be taken to conceal boats, ambulances, and other required safety equipment from the Soldier. The goal is not how much training must be curtailed due to safety precautions, but how can hazardous and realistic training be conducted safely. The safety procedures outlined in this paragraph are of a general nature. The officer in charge of the training program should use them as a guide in conjunction with the post safety regulations to integrate safety into the operation and plan for safety concurrently with other types of training preparation.

a. Air Safety.

(1) Principal instructor of problems requiring movement by, or the use of, aircraft should—

(*a*) Insure that all personnel concerned (to include assistant principal instructors (API's), Soldier and support personnel) are familiar with these responsibilities relative to flight discipline and safety.

(*b*) Provide opportunity and necessary coordination for reconnaissance of routes and target sites by aircraft personnel.

(*c*) Provide necessary control personnel and equipment for day and night operations.

(*d*) Provide necessary medical and/or evacuation personnel and equipment.

(*e*) Provide necessary firefighting equipment.

(*f*) Insure preparation of required loading manifests.

(2) Pilots.

(*a*) Final authority on each aircraft rests with the pilot.

(*b*) Prior to flight, pilot will explain emergency signals, abandon aircraft procedure, the operation of, and the signal for, opening emergency exit, and the wearing of the parachute.

(3) Safety equipment.

(*a*) Safety belts will be fastened during takeoff, landings, and turbulent weather. They will be unfastened only when authorized by pilot.

(*b*) Parachutes will be lifted or carried by the main lift webs only. Make sure pins and cones on the rear of the cover are straight and free of corrosion. Never use the parachute as a cushion or pillow. Do not place it on damp ground or allow it to come in contact with oil or grease. Adjust prior to takeoff.

(4) Conduct of troops and vehicles on airfields and/or landing zones—

(*a*) A designated control center will coordinate and authorize movement.

(*b*) After movement is authorized, movement will be performed on the double.

(*c*) Vehicles will cross runway in low gear after receiving permission to cross.

(*d*) On problems involving an airdrop of supplies, personnel will be required to move well away from drop site even if it interferes with patrol leader's plan.

(*e*) No smoking is permitted within fifty feet of an aircraft on the ground. (Smoking in aircraft will be authorized only by the pilot.)

(*f*) Remain away from arc of propeller at all times even when aircraft engine is not running and do not walk in front of aircraft when engine is running.

(*g*) Do not move about unnecessarily while aircraft is in motion.

(*h*) Remain seated after aircraft is landed until aircraft is parked and order is given.

(5) Actions to take in an emergency.

(*a*) Standby for emergency landing—fasten safety belts.

(*b*) Lighten ship—throw out equipment as ordered by the pilot.

(*c*) Abandon ship—exit immediately.

(*d*) Prepare for crash landing—as directed by pilot or his representative.

b. *Rope Safety*.

(1) *Types and characteristics of rope*.

(*a*) Nylon, 7/16-inch in diameter. Tensile strength; approximately 4,000 pounds when new. Will stretch approximately 1/3 its length.

(*b*) Manila sling rope, 1/4-inch or more in diameter. Tensile strength; should have a minimum tensile strength of 600 pounds when new.

(*c*) Three-quarter and one inch Manila ropes. Tensile strength: approximately 5,000 pounds when new. Loses much of its stretch and strength after being subjected to a heavy load, as in a fall or by being stretched too tight in a horizontal hauling line.

(2) *Care of ropes*.

(*a*) Ropes should not be stepped on or dragged on the ground.

(*b*) Ropes should not be in contact with sharp corners or edges when being used.

(*c*) Ropes should be kept as dry as possible. Do not store while damp.
(*d*) A climbing rope should never be spliced.
(*e*) Knots weaken ropes.
(*f*) Ropes weaken with age, weathering, and rough treatment.
(*g*) Acids of a corrosive nature destroy ropes.

(3) *Inspection of ropes.*

(*a*) Prior to the use of ropes for instruction or demonstrations, all ropes and items to be utilized will be inspected by the principal instructor and an officer not involved in the instruction or demonstration. This inspection will include but will not be confined to the following:

1. All ropes to be used.
2. All pulleys.
3. All pitons.
4. All anchor points.
5. All snaplinks to include those installed.

(*b*) All ropes or other items which show signs of excessive wear, fraying, rot, mildew, or other damage which might conceivably result in an accident will be separated and turned in. Equipment will be tagged to indicate the defect. If the defect is of an unusual nature, the possible cause, date, and name of the inspector will be reported.

(*c*) Normal reporting procedures for defective equipment will be followed if equipment has a defect of manufacture.

c. Water Safety.

(1) Principal instructors of problems which require the crossing of bodies of water 1.5 meters or more in depth or movement in small boats across water will—

(*a*) Inform the lane instructor of this fact and make certain that all are familiar with proper procedures.

(*b*) Remind the lane instructors and/or control personnel of their responsibilities as Safety Officer or NCO.

(*c*) Have supply issue sufficient ½-inch or ¾-inch rope to to effect crossing or sufficient small boats to accomplish movement.

(*d*) Take special disposition with regard to weak or non-swimmers.

(*e*) Erect a safety rope and place rubber boats with power lanterns and lifeguards downstream from the crossing site when deemed necessary.

(*f*) Insure power boats used for larger body of water, have radio contact with the PI and LI and contain three men (boat operator, lifeguard, and aidman) with rescue equipment.

(*g*) Assume responsibilities of Safety Officer for instruction on beaches or rivers and provide sufficient lifeguards, aidmen, and power boats for safety and/or treatment.

(2) Lane instructors will—

(*a*) Be familiar with stream crossing and small boat safety procedures and require compliance without interfering with the patrol leader's plan or the tactical situation.

(*b*) Not allow weak swimmers to carry extra equipment while crossing streams or bodies of water (radios, LMG's, etc.) and will allow no swimmer to attempt to carry more than 25 pounds of material other than the soldier's individual clothing and equipment.

(*c*) Insure that special ropes are carried by the patrol, as needed, and that individuals carry survival ropes. Inspect ropes and boats prior to use.

(*d*) Insure that equipment is properly secured for movement of boats across water.

(*e*) Assume responsibility as Safety Officer for his patrol.

(*f*) Insure that extra equipment, when carried is secured so that it cannot slip and throttle or entangle the soldier.

(3) SOP for stream crossing—

(*a*) Two Soldiers who are strong swimmers will be designated by the patrol leader to act as lifeguards. Another strong swimmer will be designated to assist the lifeguards. These men are numbered 1, 2, and 3; 1 and 2 being lifeguards, number 3 the assistant. The lifeguards should be equipped with an under arm type life preserver if available or should be provided with some buoyant material as a field expediency.

(*b*) Numbers 1 and 2 will strip down to their shorts. Their clothing and equipment will be divided equally among the other Soldiers to be carried across the river.

(*c*) The patrol leader will use the rope provided or, in its absence, will collect and connect enough survival ropes to reach across the river. Number 1 lifeguard will take up a position 10 meters down stream of crossing. Number 2 will then tie one end of small rope around his waist using a quick release, nonslip knot. Number 2 will then enter the water slowly (not dive), check the water for snags or floating debris, and swim across to the far bank towing the rope. As number 2 swims, number 3 will feed out the rope keeping it untangled and free of snags.

(*d*) When number 2 lifeguard has reached the far bank, he will untie the rope from around his waist. Number 3 will then tie his end of the small rope to the larger rope. Number 2 will then haul the large rope across the stream and fasten it to a tree 1.5 to 3 meters back from the bank. Number 2 will then take his position 10 meters down stream opposite the number 1 lifeguard.

(*e*) The Soldier will carry the equipment fixed to his harness and cartridge belt. The cartridge belt will not be fastened. The rifle will be worn slung over either shoulder with the sling tied to the harness by thongs or string with quick release knots. If an emergency arises, the Soldier should be able to drop his harness, rifle, and equipment in one quick move as though taking off a vest. All equipment must be secured so that it can be removed instantly. The Soldiers will cross the rope sliding both hands on the rope, hand-to-hand. Both hands will be kept on the rope at all times. When crossing, the Soldiers will face up stream to watch for floating debris. Too many men on the rope at once may cause the rope to sink below the surface or break. The number of Soldiers on the rope at one time will not exceed the number of lifeguards. If at all possible each man should be secured to the crossing rope by tying a short utility rope round his body and attaching it to the crossing rope with a snap link.

(*f*) Heavy equipment (non-buoyant material weighing 25 pounds or more) will be secured to the safety line in a manner that will enable it to be pushed or pulled across as the Soldier crosses. The lifeguard's equipment will be assigned to other Soldiers to carry across the stream.

(*g*) When everyone has crossed the river, number 1 lifeguard will untie the rope from the tree, tie it around his waist in a quick release, non-slip knot and swim across the river to join the other Soldier. Number 3 man will pull in the slack rope as number 1 man swims toward him. Number 2 man will remain posted as lifeguard when number 1 is crossing.

(*h*) Emphasis will be placed on preventing the necessity for using physical contact with the victim in rescue measures. Preferred lifesaving techniques is to extend some object, e.g., a stick, shirt, belt, rope, etc., to be extended to the victim as the best method in absence of a rescue boat.

(*i*) After all Soldiers, instructors, and lifeguards have crossed the river the two lifeguards will dress and secure their equipment.

(*j*) If Soldiers choose to disrobe and carry their clothing across, float equipment on brush rafts, etc., such variations will be allowed. IN ALL CASES, THE ROPE WILL

BE PLACED AND LIFEGUARD POSTED AS OUTLINED IN (*a*) THROUGH (*i*) ABOVE. SOLDIERS WILL CROSS ON THE ROPE AS OUTLINED ABOVE, EITHER TOWING OR CARRYING EQUIPMENT. Instructors will use their judgment as to whether the variation chosen is safe and will modify the Soldiers' procedure if deemed necessary.

(4) *Small boat safety precautions.*

(*a*) Without life preservers, all Soldier equipment will be attached to harness and cartridge belts and the belts unfastened. Rifle will be slung with sling loose across the body; fast release catch of harness is located so that it can be quickly operated.

(*b*) With life preservers, equipment will be attached to cartridge belt and harness. Life preserver will be put on over the harness. Rifles will be slung across the body with sling loose and quick release catch located where it can be quickly opened, not to exceed buoyance of life preserver.

(*c*) Other equipment is securely tied down in the boat.

(*d*) Instructors will be responsible for seeing that Soldiers use safe procedures in embarking and debarking from boats. They will determine who are weak swimmers and place them in boats with stronger swimmers so that help will be near in the event of an accident.

(*e*) Miscellaneous safety procedures.

1. There will be a Safety Officer or NCO at each rope site during cliff assaults (lane instructors or other persons taking part in the problem). There will be only one individual on rope at one time. Two tugs on rope is signal that rope is clear.

2. All personnel will be oriented on preventing forest fires.

3. Simulator hand granades will not be thrown any closer than 4.5 meters to an individual. Individual throwing grenades must see spot where grenade will land.

4. Machineguns with blank firing adapters attached will not be fired any closer than 42 meters to any individual.

5. Rifles and automatic rifles using blank adapters will not be fired directly at any individual closer than 15 meters.

6. Soldiers and aggressors (or control personnel) will not engage in body contact.

7. In the event of an emergency, the lane instructor will utilize the patrol's radio to contact one of the roving control points for assistance. In the event of radio failure, the injured Soldier will be moved to the nearest road and a fire built as an emergency signal. Each problem should contain a medical evacuation plan.

8. Battery or other power spotlights will be available for emergency use on night movements across bodies of water. Power boats will have spotlights while operating during darkness.

CHAPTER 4
RANGER OPERATIONS

Section I. INTRODUCTION

31. General

Ranger operations as defined in the foreword of this manual will utilize maximum available support and means of delivery. This chapter discusses certain Ranger operations that are or will be common during combat. Airmobile, ambush and roadblock, cliff assault, extended patrol, small unit waterborne, and antiguerrilla operations are Ranger operations within the capabilities of infantry units. Special training, coordinated support, and detailed planning should be considered before units attempt to conduct these operations.

Section II. AIRMOBILE OPERATIONS

32. General

a. The maximum use of all available support is considered in the planning phase of every Ranger operation. A major type of support that should always be considered is Army aircraft (fixed-wing and helicopter). Often, the very nature of a Ranger mission dictates the use of Army aircraft.

b. Due to their unique capabilities, these aircraft can be used in almost any tactical situation; examples: movement to contact, offensive, defensive and retrograde operations, patrols, and special operations behind enemy lines. The terrain and extreme weather conditions which characterize mountain, desert, jungle, and Northern operations retard movement and place a great deal of strain on equipment and men. The use of tactical transport aircraft, particularly helicopters, facilitates movement and reduces the time that equipment and troops are exposed to these stresses. Included within the capabilities of the aircraft are reconnaissance, security, target acquisition, fire support, supply and resupply, evacuation of wounded, movement of troops within the battle area, psychological warfare, and CBR.

c. The decision to use tactical transport aircraft depends largely on the type of operation, the nature of the terrain in the area, weather, and the tactical situation. Once the decision is made, the aviation unit supporting the operation is normally placed under the operational control of the supported unit commander.

d. If the operation is so large and complex as to warrant personnel being used for the purpose of terminal guidance only, pathfinders should be requested from a higher headquarters. These pathfinders are specially trained Army personnel whose primary mission is to aid in the navigation and control of tactical transport aircraft in the objective area. The inherently small size Ranger operation does not normally require pathfinder support. A brief training phase is usually sufficient to provide small units with a self-sufficient unit pathfinder capability. See FM 57–38.

e. See TM 57–210 for the characteristics, concept, missions, capabilities, and limitations of Army aircraft.

33. Operations Against Infiltration and Guerrilla Action

a. Air-landed or parachute forces are particularly suited to operations against enemy infiltrators and guerrillas. In daytime, aerial reconnaissance is used to locate infiltrators and guerrillas. Once located, Army aircraft are used to air-land support and resupply the combat patrols being used to counter this type of enemy activity. During periods of limited visibility, tactical transport aircraft position and support outposts and patrols, especially in difficult terrain likely to be used by infiltrators.

b. Small numbers of air-landed or parachute troops can patrol extensive areas; airmobile reserve units can be used effectively to destroy guerrilla bases of operation. They can also be deployed rapidly to reinforce friendly installations or units under attack by large guerrilla units.

34. Night Operations

a. Advantages. Air-landed forces may be employed effectively at night. They are less vulnerable to enemy ground and air fires and the enemy has greater difficulty in determining the location of the main landing than in daylight operations. Small air transported units landing simultaneously at widely separated points may block movement, disrupt communications, and create general confusion to assist other ground or air-landed operations.

b. Disadvantages and Problems. Night operations present certain disadvantages and special problems in comparison to daylight operations.

(1) Both ground units and tactical transport aviation units require a higher degree of training.

(2) In selecting landing zones, greater stress is given to characteristics that assist landing than to placing units on or adjacent to objectives.

(3) Ground units normally assemble after landing before proceeding on their missions, so assembly aids may be necessary.

(4) Pathfinders at landing zones and sites and special aids to navigation are more important for movement and landing than in daylight.

35. Raid

The planning for a raid is similar to that for other tactical missions. The recovery plan should provide for the transportation of prisoners and captured materiel out of the objective area. If aircraft are to be used for the recovery, this is planned prior to the operation. The aircraft may remain in the objective area to facilitate transportation during the raid or to wait for the recovery. The decision to have the aircraft remain in the objective area is based on the tactical situation, the nature of the operation, its duration, and the radius of action for the aircraft (figuring full loads for delivery and return). Recovery landing site(s) may be close to the objective(s) enabling almost immediate evacuation; however, the raiding force may divide into small groups to rendezvous with the aircraft at a predesignated point some distance from the objective(s). Alternate recovery and rendezvous points should be designated to facilitate the recovery in the event that enemy action precludes use of primary recovery or rendezvous points.

36. Patrolling

The considerations involving the use of tactical transport aircraft with reconnaissance and security forces are applicable to patrolling. For reconnaissance patrols deep behind enemy lines, the following additional factors must be considered:

a. In the planning phase, high performance reconnaissance aircraft may be used to gain information of the enemy and the terrain in the vicinity of the objective(s).

b. Aircraft will not normally remain with air-landed patrols. However, if the mission is of short duration and requires retention of aircraft in the objective area, careful consideration must be given to the security of the aircraft.

c. After the patrol has been delivered in the objective area, plans must be made for the aircraft to return to a designated place at a designated time for recovery of the patrol.

d. When the patrol is air-landed in the objective area at night or during periods of reduced visibility, the operation may require the assistance of pathfinders. For other techniques of conducting night air-landing operations refer to FM 31–20.

e. When deep reconnaissance patrols are planned, secrecy should be insured by moving to the objective area during periods of reduced visibility.

37. Operations in Special Areas

a. Mountain. Transport helicopters can place a security element on critical heights and at distances unattainable by troops moving on the ground. Helicopters, within altitude limitations, are valuable as prime movers for direct fire support weapons because of the ease with which these weapons can be moved to dominating terrain. Army observation aircraft can be used to provide observation over wide areas forward of friendly forces and to perform surveillance missions between friendly strong points. "Dead spaces" in radio reception may be overcome by establishing aircraft relay stations and aircraft may also be used to lay wire over otherwise inaccessible terrain. Additional considerations are as follows:

(1) Due to inadequate road nets, logistical support may be completely dependent upon Army aircraft. By use of Army aircraft, resupply and evacuation installation sites can be at greater distances from the line of contact and supplies can be placed closer to the location where they are needed.

(2) When formulating plans for operations in the mountains, the possibility of sudden weather changes should be considered. Alternate plans are prepared and alternate positions for air-transported forces are selected in the event the primary positions become unattainable.

(3) Single or multiple landing zones may be required for an airmobile operation depending on the enemy situation, mission, size of attacking force, and terrain. Factors to be considered in selecting landing sites are—

(*a*) The direction of wind drafts, snow, and ice covered slopes which may require troops and/or cargo to be unloaded while the helicopters hover.

(*b*) Terrain characteristics of each site.

(*c*) The necessity for pathfinders and navigational aids to marks routes and landing sites for safe movement and night landings.

(*d*) Approach and return routes which are selected to take full advantage of the defilade and concealment afforded by mountains.

(4) The actions of small semi-independent units in seizing or defending heights which dominate flight routes are of increased importance.

(5) Special clothing for personnel and equipment for aircraft must be specified and issued prior to the operation. Troops must be thoroughly familiar with the use and maintenance of these special items.

(6) The techniques of recovering air-delivered supplies and the selection, preparation, and operation of landing sites, drop-

zones, and loading sites must be understood by participating units.

b. Jungle.

(1) When formulating plans for a jungle operation, emphasis is placed on providing mobility through the employment of Army aviation. The use of aircraft in this type operation presents these considerations—

(*a*) The capability of airmobile reconnaissance and security forces to provide surveillance over large areas permits the selection of dispersed objectives.

(*b*) Because of the dense vegetation in jungles, observed fire support is limited. However, aerial observation is better than ground observation.

(*c*) The range of air-landed operations should be short if a linkup is to be made early.

(*d*) As in mountains, suitable landing sites are few in number and a shuttle system utilizing helicopters may have to be employed. Landing sites should be selected close to objectives to take advantage of the concealment afforded by the jungle and to reduce the distance necessary for movement on foot through the dense undergrowth.

(*e*) Waterways provide a means of communications and are an aid to navigation; these factors should also be considered in the selection of objectives.

(2) Supply and evacuation may be completely dependent on Army aviation. Fewer resupply and evacuation installations are required in the forward areas when forces are supported by tactical transport helicopters. Prescribed loads are determined and supplies are palletized when necessary to facilitate loading and unloading.

(3) These techniques should be considered for jungle operations—

(*a*) Troops can descend from hovering helicopters by ropes, rope ladders, or by parachute.

(*b*) Loads can be released while the helicopters are hovering over a drop site.

(*c*) Smoke or panels in trees can be used to mark landing or drop sites.

c. Northern.

(1) Enemy installations and key terrain features dominating enemy routes of communication and supply are appropriate objectives for special operations in Northern regions. The use of Army aircraft presents these considerations—

(*a*) Air-landed advance and flank guards avoid much of the fatigue caused by foot movement in snow.

(*b*) Use of aircraft enables reconnaissance and security forces to survey wide areas during daylight.

(c) Air-landed or parachute task organizations are small and compact for arctic operations. Fire support weapons are placed close to or within the objective area to avoid the difficulties of overland movement.

(d) Strong winds and blowing snow may interfere with or prevent use of aircraft: alternate plans for operating without them should be made.

(e) The brief period of daylight and the Aurora Borealis (Northern lights), which interfere with or prevent radio communications, should be considered in timing the operation. Alternate means of communications should be planned.

(f) Suitable landing zones are normally plentiful except in mountainous areas.

(g) Flight routes should take advantage of the defilade and concealment of any rough terrain in the area.

(h) Operations logistically supported by aircraft can be executed at far greater ranges than those supported by ground means.

(2) These techniques are appropriate for Northern operations:

(a) Landing sites can be marked so that they will be recognized when terrain features are blanketed or their appearance is changed by heavy snowfall.

(b) Navigational aids must be planned to overcome the effect of "white-out" (loss of reference, due to the skyline merging with the snow covered terrain).

(c) Shelters must be provided for maintenance.

(d) Portable homing devices should be provided since some areas are not mapped and recognizable checkpoints are few.

(e) Army aircraft must be completely winterized and preheating may be necessary.

(f) It is desirable that pilots be trained for ski, wheeled, and pontoon landings. During the summer months, pontoons permit utilization of the many lakes and streams as landing zones. During the breakup and freezeup periods, neither ski nor pontoons are useable on fixed-wing aircraft. Therefore, operations are restricted to wheeled aircraft utilizing available landing and takeoff areas.

d. Desert.

(1) The highly mobile reconnaissance and security units necessary in desert operations can be provided by making them air-transported. Satisfactory navigational aids are required for employment in night operations. Other considerations when using transport aircraft in desert operations are—

(*a*) Unobstructed landing zones large enough for mass landings are plentiful.

(*b*) Mountainous areas may offer the only concealment for approach and return routes.

(*c*) Aircraft can resupply and evacuate combat, reconnaissance, and security forces.

(2) These techniques are appropriate for desert operations—

(*a*) Oil or a similar substance can be used on landing and loading sites to minimize the operational difficulties caused by dust and sand.

(*b*) Pathfinders can use navigational aids such as smoke, panels, lights, and electronic devices on routes and landing zones to overcome the difficulty of terrain orientation.

Section III. AMBUSH AND ROADBLOCK OPERATIONS

38. General

An ambush is a surprise attack from a concealed position upon an unsuspecting moving or temporarily halted enemy. It is a major operational method of attack for guerrillas, but can be employed successfully by an infantry unit. The ambush can be used almost anywhere in the combat zone. Operations against guerrillas and infiltrators in the rear area may utilize ambush techniques. Forward units may employ an ambush to kill or capture enemy patrols. Patrols may be sent into enemy territory with the mission of ambushing and destroying an enemy vehicular convoy, train, or killing or capturing troops of a foot column. The key to success in the conduct of an ambush is the element of *surprise.*

39. Conduct of an Ambush

The ambush is characterized by ruthless and violent action. Proper intelligence, planning, and coordination are necessary when formulating plans for this operation. See FM 21–75. Factors to be considered are—

a. Selection of the Ambush Site. When selecting the ambush site, a careful study must be made using maps, aerial photographs, and when possible, a personal reconnaissance. Consideration must be given to the availability of natural obstacles that can be used. A good example of this is a route with a steep cliff or swamp on one flank and good positions from which to deliver a heavy volume of fire on the other flank. The absence of natural obstacles can be overcome by substituting minefields and road craters or establishing killing zones with the Claymore antipersonnel weapon. In a successful ambush, the enemy is cut off and unable to deploy his forces. Tricks or ruses to lure him into the ambush should be considered. Road signs can be placed to confuse enemy columns. Feints or raids can be made at one site in order to direct the unit to the desired location for ambush. Commanding ground, existing cover, and concealment should be fully utilized.

b. Route to and From the Ambush Site. The route to and from the ambush site is carefully selected to insure secrecy in occupying the positions and speed and security when withdrawing from it. Maximum cover and concealment are desirable. After entry into the ambush area, the route followed by the ambushing force is carefully inspected to remove all evidence of the force's presence. In planning the withdrawal, alternate routes are selected to avoid the possibility of enemy forces blocking the withdrawal of the ambush force. These routes are patrolled periodically to insure that enemy infiltrators do not occupy them.

c. Communications and Control. Control is necessary during the movement to, occupation of, and withdrawal from the ambush site. The most crucial time of the ambush operation is the moment the enemy arrives at the site. Comunications must provide for the issuance or orders to open fire. If this is not possible, the time the enemy's lead element arives at a certain location is designated at the time to open fire. Communication with local security elements and higher headquarters is desirable. Exacting control must be exercised to insure that the ambushing force is alert and silent. Once the force is in position, movement must be kept to a minimum. Care is taken to insure that the ambush is not executed prematurely. Assembly and rallying points are designated to assist in control during the withdrawal.

d. Rehearsal of Participating Troops. Rehearsals for the ambush operation are conducted on terrain similar to that which is to be used on the actual mission and must include all personnel scheduled to participate in the action. Subordinate leaders, weapons crews, and security elements are briefed until all personnel know the exact sequence of events of the ambush and thoroughly understand their duties. Equipment is checked and all weapons are test fired at the rehearsal. All men should know the routes of withdrawal and location of assembly points.

e. Camouflage Measures. In no other operation is camouflage discipline more important than in the ambush. Personnel and weapons must blend with the surrounding area as much as possible and all residue resulting from preparation of the site must be removed. Once the position is camouflaged, personnel must not move unnecessarily. Patience and staying power is a must to avoid premature disclosure of the ambush.

f. Coordinating Fire Plan. The fires of all weapons, including long-range artillery, close-in automatic rifles, rocket launcher, grenade launchers, light antitank weapons, Claymore, and other weapons are tied into the fire plan. The time to open and cease fire, the assignment of sectors of fire, and the location of friendly forces are considered. Plans are made for isolating the ambush area to prevent escape and reinforcement of the enemy. The effectiveness of the ambush relies upon the *surprise* delivery of a large volume of fire. The fire should come from at least two directions and converge on the target. Care is taken to prevent friendly troops from firing into other positions of the ambush area when converging fire is used. To achieve surprise, fires of the ambush force do not commence except upon a prearranged signal. Fire support covering the withdrawal is considered in fire support planning and coordination.

g. Use of Assault Element. The use of "killer teams" in the assault element in conjunction with the ambush is often desirable. The purpose of these teams is to physically move through the ambush site and assure the destruction of vehicles and material, search enemy dead, or accomplish any other duties considered necessary by the ambush commander. The action of this team must be planned. rehearsed in detail, and violently executed. The nature of the ambush usually dictates the strength and employment of this team.

h. Use of Support Element. The formation of a separate support element is often desirable in ambush operations due to the maximum use made of fire power. The assault element fire power complements the support element and may sweep the objective in a killer role after the lifting or shifting of fires.

i. Reasons for Failure. Some of the primary reasons for failure of ambush operations are—

(1) Disclosure of ambush site by noise of weapons.
(2) Tendency of men to shoot high.
(3) Disclosure of ambush position by footprints made moving into position and movement by individuals when enemy is approaching.
(4) Lack of fire control. Commanders were unable to stop firing and enter next phase of plan.
(5) Patrol leader not in position to control his element.
(6) Lack of all-round security to include security teams.
(7) Failures in weapons due to dirt, failure to inspect and test fire.
(8) Lack of clearly defined signals for ambush plan.
(9) Poor target area responsibility causing fire to be wasted on same area.
(10) Premature firing.

40. Defense Against Ambush

In planning for defense against ambush, the planner must initially consider the friendly force available. A large vehicular column reinforced with armor reacts differently against attack than a small unit of foot troops without reinforcements. The small unit commander responsible for moving a unit independently through areas where ambush is likely must plan for: The formation to be used; march security; communication and control; special equipment; the action to be taken if ambushed; and the reorganization.

a. Formation. A dismounted unit normally uses a formation that provides all-round security while en route. Responsibility for this is assigned to subordinate commanders. The distance between units in march depends on the terrain, visibility, and situation. March interval is also great enough to allow each succeeding element to deploy when contact with the enemy is made. However, the distances are not so great as to prevent each element from rapidly assisting the element in front of it. The column commander should be located well forward in the formation, but is not restricted from moving throughout the formation as the situation demands. Units are placed in the formation so they may distribute their firepower evenly throughout the formation. If troops are to be motorized, unit integrity is maintained when possible.

b. March Security. Regardless of whether the unit is on foot or motorized, security to the front, rear, and flanks is necessary when ambush is likely. A front security element is placed well forward with adequate communications with the main body. The security element is strong enough to sustain itself until followup units can be deployed to assist in reducing the ambush. However, the enemy may allow a security element to pass in order to attack the main body. If this occurs, the security element may be used to attack the ambush position from the flanks or rear in conjunction with the main action to destroy it. Flank security elements are placed out on terrain features adjacent to the route of march. They move forward either by alternate or successive bounds, if the terrain permits. This is often difficult because of the ruggedness of the terrain and the lack of transportation or communications. The next best thing is moving adjacent to the column along routes paralleling the direction of march. Rear security is handled similarly to front security and plans can be made for the rear guard to assist in reducing the ambush either by envelopment or by furnishing supporting fire. Reconnaissance by fire of likely locations for ambush may greatly assist the security forces; however, extreme care must be exercised by the convoy commander when authorizing this expedient because of the possibility of unknown friendly units in the area. Light aircraft, flying above the column on reconnaissance and surveillance missions, increase column security. When available, fighter aircraft can provide column cover. Air to ground communication between these elements is highly desirable.

c. Communication and Control. All available means of communication are used, consistent with security, to assist in maintaining control of a small unit during movement when ambush is likely. March objectives, checkpoints, and phase lines may be used to assist the commander in controlling his unit. Communication with security elements is mandatory. Detailed prior planning, briefing, and control pays off if an ambush does occur. Alternate plans are made to cover all possible situations. Higher headquarters is notified as soon as possible of the ambush so it may alert other units in the vicinity.

d. Special Equipment. It is often necessary to provide the unit with additional items of equipment and weapons, especially when it moves through areas where guerrillas or other enemy forces are likely to be encountered. It is desirable that vehicles have an automatic weapon mounted and manned. Pioneer tools and mine detectors are used to detect and reduce roadblocks and minefields. Demolition equipment is used to destroy obstacles encountered en route. Additional repair parts may be needed to make "on the spot" repair of disabled vehicles. Ample communications equipment is always necessary, including identification devices to friendly aircraft such as panel sets, lights, or smoke grenades.

e. Actions to be Taken if Ambushed. The most effective means of combating an ambush is through the immediate return of fire by all weapons. This requires discipline, dynamic leadership, and rehearsed plans. Regardless of the methods of movement, all weapons should be positioned for immediate use. A well placed and smoothly executed ambush is difficult to counteract. However, even the effectiveness of a well planned ambush can be reduced by this immediate return of fire. Critical positions are often hit and important weapons silenced, thus creating weak points in the ambush position. Personnel must be prepared to immediately exploit such conditions by assaulting these weak points. Smoke and grenades are extremely effective in executing counterambush actions. They will create confusion, offer a screen for movement, and disrupt the ambush plan. A unit must be made aware that an assault (front, flank, or rear) on the enemy position takes place despite the fact that the unit may have suffered heavy casualties. This assault must be made to prevent complete annihilation of the friendly unit. A retrograde action in lieu of the assault will, in a well prepared ambush, result in complete defeat for the friendly unit. When all elements of a force are not trapped within the ambush site, the elements that are free to maneuver will initiate an immediate flank or rear assault against the ambush force. A flank assault will permit better coordination between the ambushed and free elements. Supporting fires are limited to those weapons which cannot be handcarried and fired in the assault. Higher headquarters will specify if units successfully overcoming an ambush will form a pursuit force to destroy the fleeing ambush group. Any delay in this operation will act in favor of the enemy and permit his conducting future ambush actions in this zone. When friendly units are motorized, trucks being used as personnel carriers, the tarp and bows are normally removed. Troops riding in these trucks remain alert at all times and are trained to fire and detruck immediately when ambushed. Necessity may require that they dismount even before the trucks are fully halted. Assistant drivers are assigned to all vehicles in case the driver becomes a casualty. If an ambush occurs, drivers are instructed to halt their trucks on the road. They do not pull off onto the shoulders because the shoulders may be mined. If practicable, floors of lead trucks should be reinforced with sandbags to reduce the effect of mines.

f. Reorganization. The reorganization after an ambush involves the use of rallying points, plans for local security, reorganization, and movement based on the unit mission. Clearing minefields and other obstacles, evacuating wounded, and disposing of disabled vehicles present special problems.

g. Failure. All plans and preparations are wasted if personnel fail to be alert and vigilant at all times. This applies especially to security elements. Consequently, duties such as flank guard are rotated often.

h. Small Combat Units. The principles discussed in this section can be applied with modification to all relatively small combat units.

41. Establishing a Roadblock

a. The roadblocks discussed in this section may be either temporary (hasty) or deliberate. The temporary roadblock commonly used with an ambush is designed to halt the lead elements of the enemy force when it arrives in the ambush area and to "trigger" the ambush action. An example of a hasty roadblock is a tree wrapped with enough charges to fell it across an approach route when the enemy comes within the ambush area. Another example is an antitank mine buried in the route. The preliminary preparations for a hasty roadblock, such as placing charges or positioning vehicles, are completed as rapidly and silently as possible to keep the location, strength, and mission of the ambush secure. The deliberate (permanent) roadblock is normally used to block routes into friendly battle positions or to force the enemy into a position favorable to friendly defenses. The construction of a deliberate roadblock depends on the time, material, and personnel available to build it.

b. The most common permanent roadblocks are—

(1) *Antitank ditches.* This roadblock is usually employed in conjunction with minefields and wire entanglements. It may be dug by handtools or with bulldozers.

(2) *Side hill cuts.* This type roadblock is placed with demolitions by the employing units. When locating this road-

block, the employing unit must consider the availability of bypass routes the enemy may use.

(3) *Road craters.* A road crater is a large hole in the center of a route of approach that is designed to deny the enemy the use of the route.

(4) *Log obstacles.* Log obstacles are either rectangular or triangular in shape and are usually filled with rocks or dirt. Log posts may be dug into the ground to form a similar type obstacle.

(5) *Abatis.* An abatis is constructed by felling trees at an angle of about 45° to the enemy approach. The trees are left attached to the stumps to prevent rapid removal and may be booby trapped.

(6) *Steel and concrete obstacles.* Roadblocks of steel and concrete may be constructed when personnel, time, and material are available. Dragonteeth, I-beams, ramps, or other designs may be used. For complete details of roadblock and obstacle construction see FM 5-15 and TM 5-310.

c. To be effective, the roadblock is placed in a position to deny the enemy the opportunity to bypass it or to sufficiently deploy his forces so as to conduct a strong attack against it. It is covered by accurate fire. Full use is made of natural obstacles such as swamps, dense woods, or extremely rough terrain. Antitank and antipersonnel mines may also be used. A good roadblock should be—

(1) Placed along a likely avenue of approch.
(2) Strongly constructed.
(3) Difficult to destroy.
(4) Constructed with materials available locally.
(5) Covered by accurate fire.
(6) Concealed from enemy observation.

Section IV. CLIFF ASSAULT OPERATIONS

42. General

This section outlines the techniques of assaulting a cliff obstacle as might be found during an amphibious or waterborne raid. These techniques may, however, be used by troops encountering an obstacle of this type regardless of its location or the nature of the operation. The discussion of cliff assault techniques involves a patrol's landing on a hostile shore, its actions on the beach, assaulting and scaling of the cliff, actions on the cliffhead, and actions during the withdrawal down the cliff.

a. The element of surprise is an essential part of the successful raid. In order to increase the opportunity of surprise at the objective area, the leader of an amphibious raid carefully considers his choice of a landing place. In most cases, the easiest landing can be made on a steeply shelved sandy beach. The patrol leader should, however, consider avoiding such a landing place because it probably will be defended.

b. If all members of a raiding patrol are prepared to swim and are able to climb a rock cliff, there exists a chance of getting ashore unopposed. This unopposed landing enhances the opportunity for surprise later at the objective area. Therefore, it is often preferable to accept the disadvantage of landing upon a beach with physical difficulties such as rock cliff obstacles, rather than a relatively easy beach approach with its probability of being well defended.

43. Special Equipment

a. In addition to the equipment necessary for the conduct of the raid itself, the patrol may need special equipment for overcoming the cliff obstacle on the beach. Metal scaling ladders, climbing ropes, toggle ropes, rope ladders, grappling hooks, and "bear claws" might be used. Rockets with grappling heads may be necessary to carry the ropes over the cliffhead. It may be necessary for the first climbers to have additional mountaineering equipment such as pitons and hammers, snap links, and sling ropes.

b. Communication equipment such as sound-powered telephones may be helpful in control at the beach and cliff. Engineer tape may be used for control on the cliffhead.

44. Initial Landing

a. Landing. The landing should be in two waves. There are three parties in the first wave: the climbers who establish the scaling ropes or ladders, the patrol leader, and the beach security personnel.

b. First Wave. The first men ashore secure the boats at the beach while the others disembark. Next ashore are the climbers. For a company-size raiding party, there normally should be a total of six senior climbers who initially establish the fixed ropes. Three of these climbers are designated the No. 1 climbers and three the No. 2 climbers to form 3 teams of 2 climbers each. Once on the beach they move directly to the base of the cliff. The No. 1 man on each team then begins his ascent. The No. 2 climbers at the base of the cliff tend the No. 1's ropes belaying him in his ascent when required (fig. 11).

(1) Third element of the first wave is the beach security. These men take up defensive positions on either flank of the landing area at the foot of the cliff. One man is designated the beach control officer. With the help of a messenger and radiotelephone operator, he establishes the beach control team. The radiotelephone operator establishes the base telephone. The remainder of the first wave takes up firing positions

Figure 11. Initial landing—beach secured, No. 1 ropes established.

while the climbers establish simple fixed rope installations. If the cliff is higher than the length of a climbing rope, then two or more ropes may be tied together, or where possible, an intermediate anchor point may be established to tie off a second fixed rope. The boats withdraw as soon as they are cleared of the raiding party personnel.

(2) Upon reaching the cliff top, the No. 1 climbers secure their ropes.

(*a*) It may be necessary to use "bear claws" for this if no object is available around which the rope can be tied. Care should be taken that the rope is not obviously exposed on the cliffhead. This can be done by placing the "bear claws" under a small section of sod. A bayonet or entrenching tool can be used to remove the sod. The claws can then be placed under the sod and the spikes pushed into the ground. The rope should then be secured to the ring and the sod replaced (fig. 12).

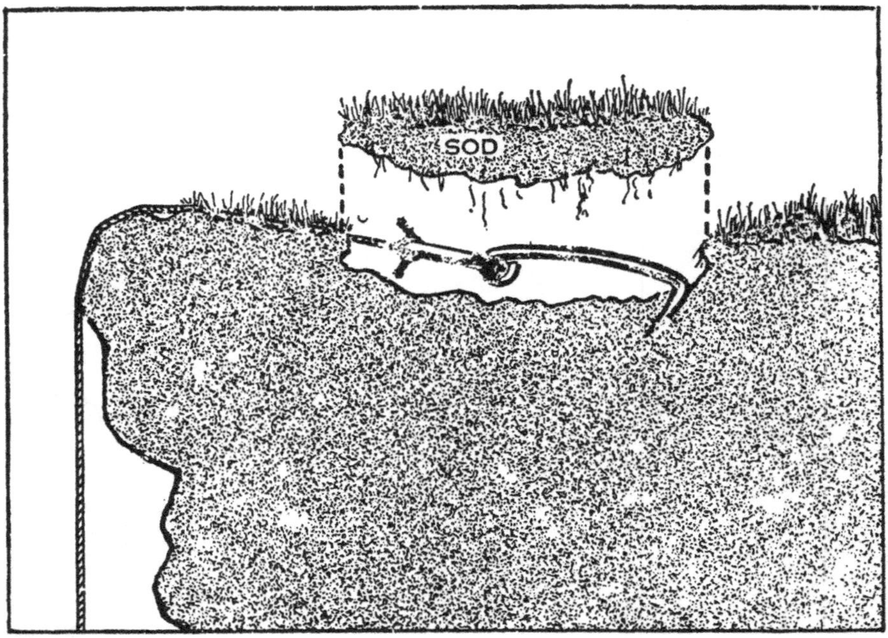

Figure 12. The rope and claws should not be exposed on the cliffhead.

(*b*) The No. 1 climbers then make a hasty reconnaissance of the cliffhead area to insure that the immediate area is unoccupied by the enemy. Having insured that the area is clear, they signal to the No. 2 climbers that they have cleared and secured the fixed ropes. This signal should be two or three tugs on the rope. Each succeeding climber uses the same signal upon reaching the top to indicate the rope is clear. There should never be more than one man on any one rope. Upon being signalled, the No. 2 climbers immediately begin their climb. They carry a spare rope and "bear claw" with them (fig. 13). On reaching the cliffhead, they secure their spare ropes so that six fixed ropes are available for climbing.

(3) As soon as the ropes are vacated by the No. 2 climbers, the patrol leader and security teams climb up the ropes. They are followed as quickly as possible by the remainder of the first wave, leaving the beach control team below (fig. 14).

(4) The second wave, the main force, is signalled in by the beach control officer. Meanwhile, the security team on the cliffhead takes up defense positions on both flanks.

(5) The patrol leader selects a control point at which the remainder of the patrol reports on reaching the top. He also selects a location for his headquarters. White engineer tape is laid from the control point to the headquarters and from

the control point to the two outer ropes in the climbing area (fig. 15). This serves to canalize the remainder of the force into a central control point as they reach the top of the cliff. From this control point they are then directed to their various positions in the defense of the cliffhead.

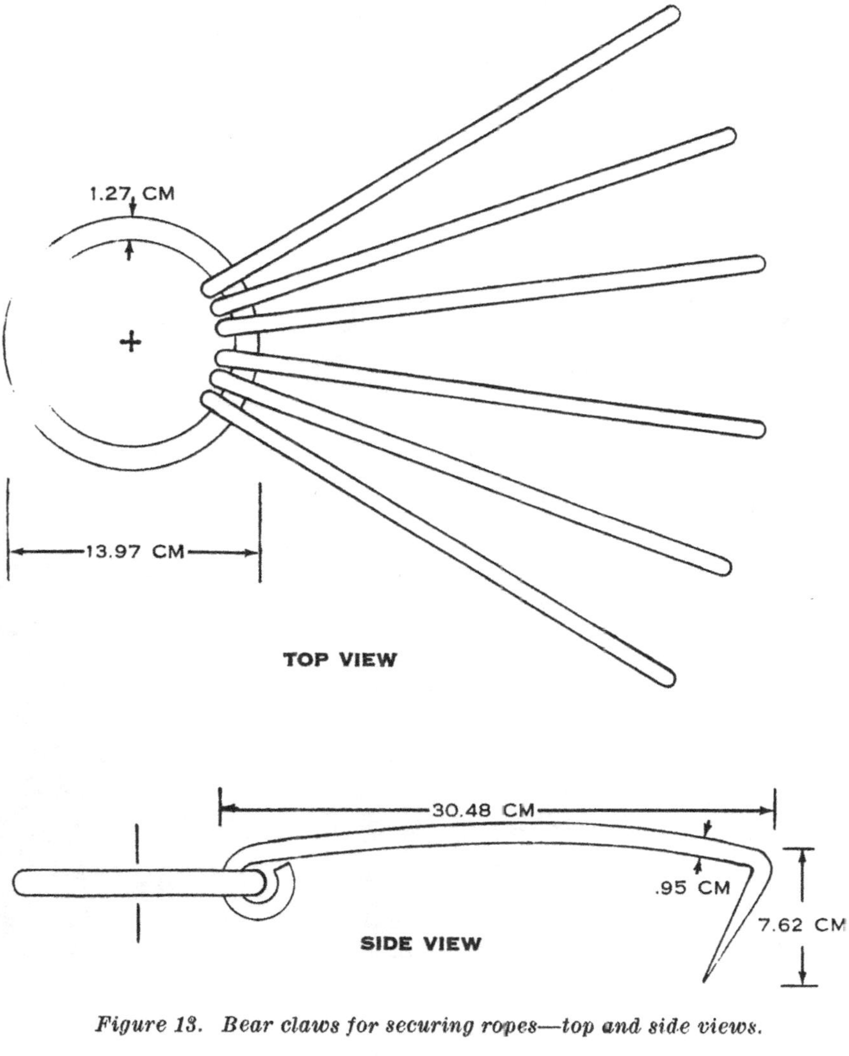

Figure 13. Bear claws for securing ropes—top and side views.

(6) One man, designated the cliffhead officer, is located at the control point to direct the others as they move in.

(7) The radiotelephone operator, with the patrol leader, establishes telephone communication with the operator in the beach control team at the base of the cliff.

c. Second Wave. The second wave initially takes up defensive positions on the beach and then moves for the ropes as directed by the beach control officer. The beach control officer insures that all ropes are in use.

(1) The first up are the subordinate leaders in the main force. Upon reaching the top, they move inland along the tapes to the control point where they are directed to their respective sections in the cliffhead defense. They lay engineer tape as they move from the control point to their sections.

Figure 14. All ropes established—cliffhead secure.

(2) The remainder of the force follows and takes up positions in their sections. As each man moves into the control point from the cliff edge, the cliffhead officer directs him to move along the tape leading to his section (fig. 16).

(3) As soon as the force is in position, a runner in each section reports to the cliffhead officer and takes in his tape on return to his section. The cliffhead officer then reports to the patrol leader that all men are in position.

(4) The patrol leader then leads the raiding party toward the objective, leaving the cliffhead officer and a detachment to defend the cliffhead.

(5) The cliffhead officer reorganizes the remainder of the force into a cliffhead team. The No. 1 and No. 2 climbers are left with this detachment. The No. 2 climbers remain at their ropes. The radiotelephone operator in the beach control team then joins the cliffhead team and establishes radio con-

Figure 15. Control on the cliffhead.

tact with the patrol leader. One of the No. 1 climbers doubles his rope in preparation for the withdrawal so that it can be retrieved from below. This may require the use of an additional rope in order to reach the base of the cliff.

(6) The cliffhead team then takes up security positions and awaits the return of the main force (fig. 17). The cliffhead team does not commit itself to the enemy if at all possible. Enemy beach sentries are allowed to pass through the area unless

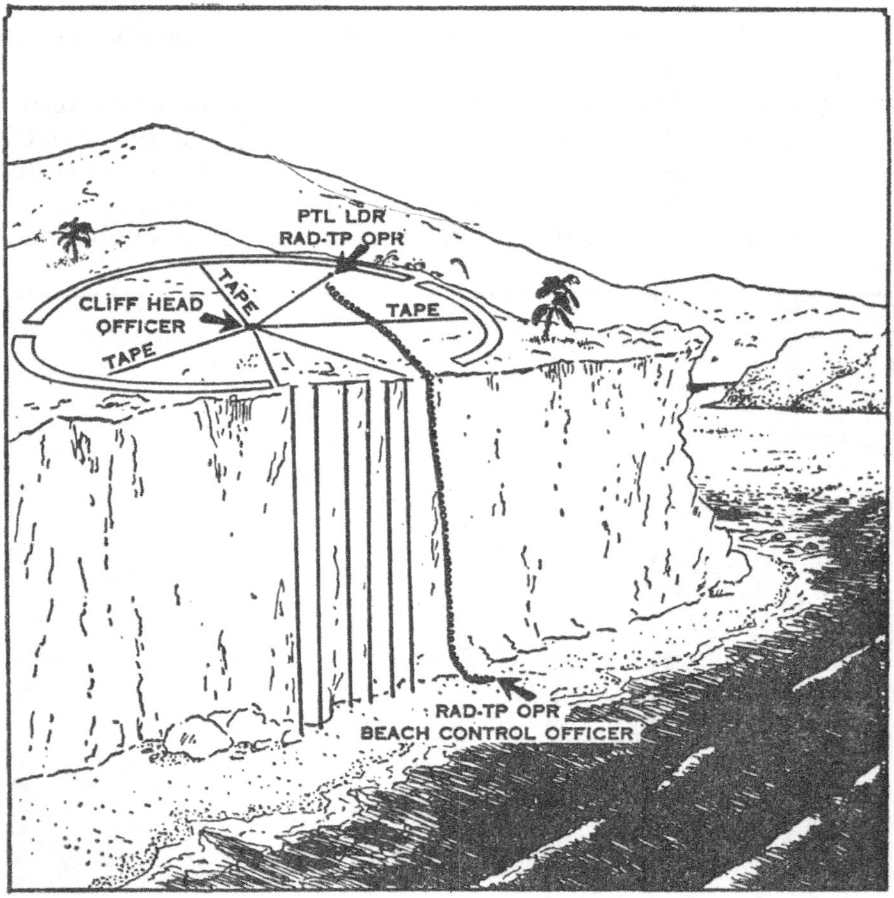

Figure 16. Organization of the cliffhead.

there is a possibility of their seeing the main force. Every precaution is taken to prevent compromising the cliffhead area to the enemy. If sentries have to be eliminated, they should be taken silently. Care should be taken to properly conceal both the enemy dead and their equipment.

45. Withdrawal

a. If the situation permits, the main force returns to its original assembly area at the cliffhead, allowing for casualty evacuation and reorganization. Tapes from the control point and headquarters to the cliff edge may be relaid if necessary. In the event of close pursuit, the cliffhead team engages the enemy while the main force evacuates.

b. The patrol leader orders the party to prepare for withdrawal. On this command, the beach control officer signals the craft to come in. The beach control officer with his radiotelephone operator establishes a control point at the foot of the cliff. The No. 2 climbers go to the

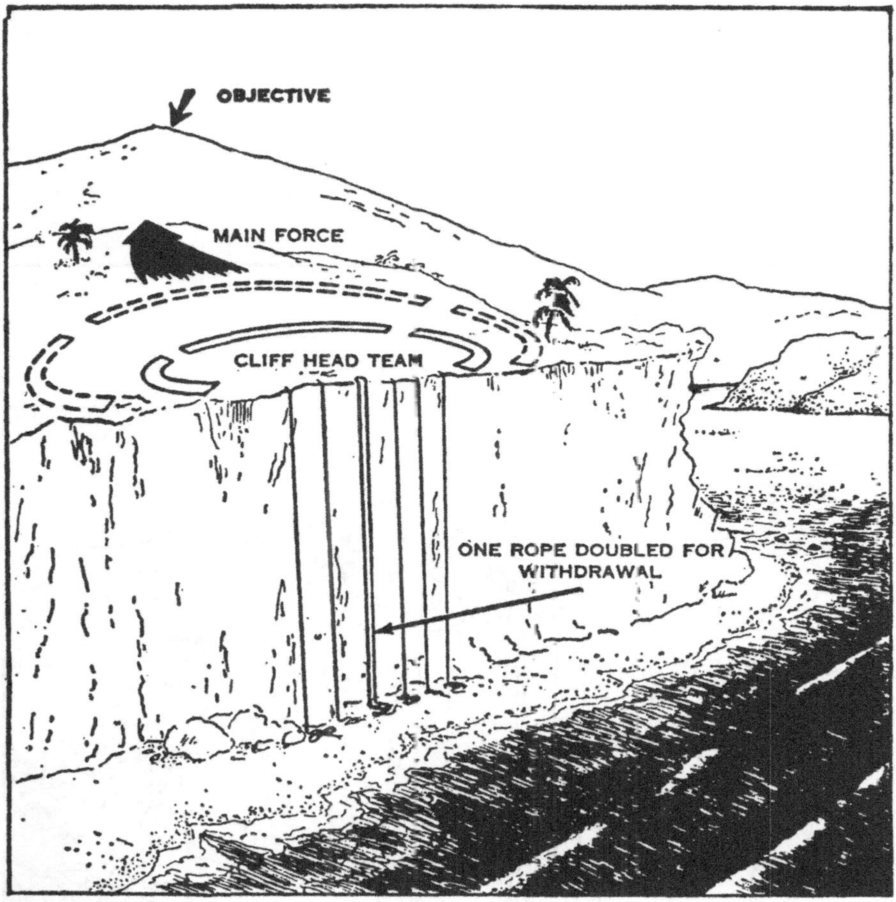

Figure 17. Reorganization of the cliffhead.

rope tops. On the order to withdraw, the subordinate leaders of the main force withdraw to their ropes and send their machineguns to join the cliffhead team. The cliffhead officer insures that all ropes are used. The subordinate leaders are the last of their sections to rappel down. They report to the beach control officer on arrival. The cliffhead team then filters down and withdraws, followed by the patrol leader. The patrol leader's radiotelephone operator takes in his telephone lines as he goes. The No. 2 climbers cast off their ropes, recover their "bear claws," and descend the No. 1 ropes (fig. 18). The No. 1 climbers release all single ropes and descend the remaining double rope (fig. 19). The last man down pulls the double rope down upon reaching the base of the cliff (fig. 20). The remainder of the force then embarks. Throughout the withdrawal, each man takes every possible precaution to prevent any of his equipment from being left either on the cliffhead or on the beach. The area should be completely void of signs or traces of the patrol's presence in that area. The patrol leader

Figure 18. No. 2 climbers release their ropes and descend the No. 1 ropes.

Figure 19. No. 1 climbers release their single ropes and descend the remaining double rope.

Figure 20. No. 1 climber descends the double rope and then pulls it down.

and all of his assistant leaders should take extra precautions in controlling this factor during the withdrawal. Regardless of whether the raid is a success in material damage to the enemy, in all certainty it will be a success against his morale. His enemy has appeared on his ground, struck a blow, and vanished completely without a trace.

Section V. EXTENDED OPERATIONS

46. Introduction

Nuclear warfare and probable enemy tactics have increased the requirement for not only standard type patrol missions of short duration, but missions requiring an extended time-distance factor. These extended operations introduce problems of education and training, of teaching self-confidence and use of improvised techniques as well as selection of individuals to conduct these operations. Extended patrol operations require firm, farsighted, and energetic leadership by commanders to cope with the demands of various situations. No two operations are exactly alike. Training depends on the type missions to be accomplished. Variances of missions will require a highly skilled Soldier not easily confused by rapid changes. These requirements imply the need for specially trained units or special training of individuals in conventional units. A well supervised, aggressive Ranger training program will fulfill most of their requirements at division level and below.

47. Purpose and Scope

This section will define types of missions for extended patrol operations and include commander's considerations for their preparation and execution. Operations will involve small units and normally be geared to foot mobility in the objective area. Their success will depend upon sound organization and thorough planning and timely execution on the part of the commander. The organization and planning will vary with each mission and operation.

48. General

a. Patrols are classified by their missions as combat or reconnaissance patrols. These patrols can further be designated as short-range or long-range patrols, depending on the time/distance factors involved in performing the mission.

b. Extended operations involve the employment of long-range patrols in most instances. However, such operations are not restricted to patrol actions and may be carried out by units or organizations.

c. Extended operations, regardless of the size or type force involved, require special considerations in regards to clothing and equipment, resupply, communication, personnel selection, training, support, organization, and planning.

49. Missions

Short-range or long-range patrols may be assigned either combat or reconnaissance missions. Typical missions include:

a. Target Acquisition. The detection, identification, and location of a target in sufficient detail to permit target analysis and effective weapons employment.

b. Area or Zone Reconnaissance. The gathering of information within a defined area. Reconnaissance by fire may be a technique used in accomplishing this type mission.

c. Route Reconnaissance. The gathering of information about a specific route or routes. Mobility often favors the use of air or vehicular transportation.

d. Point Reconnaissance. Reconnaissance of a specific location. Foot patrols within the objective area normally offer the greatest degree of success if the point is occupied by the enemy.

e. Surveillance. Continuous observation of an area, route, or point which is usually performed by small detachments having an adequate radio communications capability. This patrol gains information on enemy movements, locates nuclear and nonnuclear targets, and may be used in CBR monitoring or survey roles. Long-range operations usually favor air or sea transportation for movement to the general operational area and for resupply and evacuation.

f. Reconnaissance in Force. Normally conducted by a considerable force to discover and test the enemy's position and strength. Mobility, firepower, and communications are of primary importance in this type of operation.

g. Contact. Maintaining contact, visual, radio, or physical, with the enemy or adjacent friendly units.

h. Ambush. A surprise attack from a concealed position upon an unsuspecting moving or temporarily halted enemy.

i. Raid. A sudden attack usually by a small force having no intention of holding the area attacked.

j. Harassment. A patrol operating behind enemy lines or in close proximity to the enemy's forward elements having the mission of crippling his communications or hindering his movement.

k. Patrol of Opportunity. Operations of a patrol in a specified enemy zone with the mission of destroying targets of opportunity. Certain target limitations may be imposed by higher headquarters which, if destroyed, would hamper future friendly operations, e.g., bridges.

l. Stay-Behind. A unit or force of any size which purposely allows itself to be bypassed by the enemy to perform a mission(s) behind enemy lines. These operations may allow concealment of extensive supplies prior to the enemy bypass and offer an excellent opportunity to conduct extended guerrilla-type operations within the enemy's rear area (par. 50).

m. Mopup Operations. The mission of locating and destroying bypassed enemy elements, enemy harassing patrols, or guerrilla forces operating within friendly held territory.

n. Resupply. The resupply of dispersed units by a task-force type organization composed of logistical and combat arms. It is used when enemy and friendly forces are intermingled or when enemy patrols are active in the friendly rear area. Resupply patrols may also be used to reinforce or resupply long-range patrols, especially in Arctic, desert, jungle, or swamp operations.

o. Prisoner Seizure. The employment of raid-type tactics against enemy positions for the purpose of obtaining prisoners.

p. Liberation. The employment of raid tactics against an enemy installation for the purpose of releasing friendly prisoners. Evacuation usually favors the use of air.

50. Methods of Delivery and Withdrawal

a. Consideration is given to the method of delivering the patrol to or near the target area when the mode of transportation is not used as a fighting vehicle. Methods of delivery are dependent upon several factors which include the mission, the enemy situation, delivery means available, depth of penetration, weather, target priority, air superiority, and terrain. The most desirable method selected is normally the one which reduces the possibility of detection. Once in the area in which it is to carry out its mission, with few exceptions, the patrol becomes foot mobile in order to maintain secrecy. The importance of delivery without detection is always desirable and in specific instances mandatory for the successful accomplishment of the mission. Methods of delivery are as follows:

(1) Air delivery.

(*a*) Parachute.

(*b*) Air landed—helicopter or aircraft.

(2) Infiltration.

(*a*) Ground—generally by foot and possibly by vehicle.

(b) Amphibious—by assault boat, submarine, and other water transportation.

(*c*) Stay-behind—patrol of a withdrawing unit which remains in concealment as enemy forces gain control of the area.

b. The method of recovery is dependent upon the mission, enemy situation, terrain, means of recovery available, proximity of friendly forces, and future patrol missions. In most instances, the expected method of recovery will be known prior to the patrol's entry into the objective area; however, an alternate plan of recovery should always be considered. The headquarters initiating the patrol is normally responsible for providing its means of recovery.

51. Training and Personnel Selection

Training and personnel selection are the first considerations in organizing and conducting patrolling operations. Sufficient doctrine exists for the training of the combat arms in short-range patrol operations. The basic prerequisites include a high degree of proficiency in infantry skills. Long-range patrol missions are extremely varied in nature, requiring specialized training, mental endurance, and intestinal fortitude. Common to all extended operations is a requirement for men of excellent caliber. At army and corps level it may be advantageous to organize separate units for exclusive use in extended operations or they may select a specific unit to be specially trained to function in this capacity. Use of such units for other than extended operations should be discouraged because of the possible heavy loss of highly skilled personnel.

a. Personnel Selection.

(1) Personnel should be volunteers. This does not mean that an organization's best combat potential is taken away. Volunteer status appeals to individuals with various backgrounds and degrees of experiences. If the higher commander selects a specific unit, it should be one of his best, and it should have the same spirit as that of any group organized solely from volunteers.

(2) Physical requirements are normally higher than those of standard combat units. When molding a group of *individuals*, some attrition will occur although all appear physically fit. The problem in this case is mental endurance. The conditioning program should overcome such a problem. When working with a *unit*, the individual who is weak is forced to meet the standards being met by other members of his organization. Continued inability, however, will dictate elimination.

(3) Commissioned and noncommissioned officers must be versatile, quick-thinking, and capable of improvising as situations demand. Each must be prepared to assume responsibility or act independently. Personnel selected must be endowed with commonsense and a desire for the daring.

(4) Since the success of extended patrol operations depends largely on the mental attitude of all patrol members, an *individual* volunteer should be forced to return to his parent unit if unable to meet established standards. In the case of *unit* training, the situation is more complex. The commander should carefully examine his organization prior to selection to determine which units demonstrate the qualities necessary to accomplish such missions. When members of the unit undergoing training fail to show effective development or will prove to be a major hindrance to unit progress, no choice exists except to eliminate them by transfer to other units.

(5) Just as every man must be mentally conditioned for unit isolation on the battlefield, each member of an extended patrol unit must be mentally prepared for individual isolation. He must further be conditioned to desire close combat with the enemy and be trained in the use of silent weapons. This requires continuous training and constant indoctrination.

b. Training.

(1) *Training objective.* The primary objectives of extended operational unit training are as follows:

(*a*) To instill increased discipline and esprit de corps.

(*b*) To teach that darkness, difficult terrain, and weather extremes reduce casualties and help to insure successful accomplishment of the mission.

(*c*) To develop in each individual an extremely high level of physical fitness while teaching him the required skill in weapons and equipment which enable successful accomplishment of any mission.

(*d*) To instill in the individual a desire for contact with the enemy under all conditions.

(2) *Training scope.* The training scope is dictated by the mission assigned. In addition to special training for extended operations all patrols should have common capabilities. The patrol must be able to establish and maintain adequate communications. It is desirable that patrol members be cross-trained in all communications equipment available. Unit pathfinder and demolitions capabilities are necessary. Special capabilities for short-range patrols may be gained through the use of attached specialists. This is normally inadvisable in long-range patrols. Without training to insure proper mental and physical conditioning, specialists may impede the progress of the patrol and adversely affect the mission. The training scope for extended operations includes this specialized training as part of its overall training program. The scope should not be limited to any one type mission and should include subjects required for any mission.

(3) *Training for geographical area.*

(*a*) Desert, alpine areas, the Arctic and other geographical operational areas must be considered when training for extended operations. Units or individuals trained under desert conditions will normally fail in Arctic operations without additional training. Although the items of equipment may remain the same, maintenance requirements and expedients used differ radically. For instance, individual weapons in the Arctic must be cared for, cleaned, and lubricated differently from those in the desert. A unit may find that a specific item selected for use may be ineffective in the jungle and yet highly desirable for desert operations.

(*b*) Geographical adaptations are of two types, group training in specific subjects and the individual's adaptation to his environment. Both are of equal importance. The first is gained through formal training. However, the second is self-taught and provides the foundation for extended operations. The Soldier learns what his body is capable of doing in his new environment; through experience, he determines what techniques are best for him. As an example, the individual learns that he can walk many miles on blistered feet; with more experience he learns a technique of taping sore areas which eliminates blisters.

While another man finds this technique will not work for him, he finds that wearing two pairs of socks is more effective. Each individual learns his own salt and water requirements for the geographical area. The group learns the best expedient for maintaining group proficiency in each zone. For instance, it finds that group shelters are better suited for the Arctic, even though this technique may be unsuitable in other areas. Each training day in an operational zone increases the extended operational ability of the unit.

(c) The training requirements imposed by geographical area are not detrimental time-wise. Extended operations require an endurance and acclimation factor over and above normal unit operations. A patrol of opportunity may lose its potential when all effort must be expended to survive. When the expedients are learned, they become habit and permit the special type of operations to be effected. The key to rapid environmental adjustment is the initial individual selection previously discussed. Units which set adequate selection standards can be operational within 30 days after arrival in a geographical area. Extended operational units located in the ZI can maintain their operational efficiency in four geographical areas during their annual training cycle and need not require an adjustment period after arrival in a combat zone.

(4) *Foot movement training.* Training in foot movement is an important consideration because it offers the greatest form of mobility. It must be included in the training program. Training is progressive in time-weight-distance factors. Care is taken to gradually increase the individual's load, rate of march, and distance traveled.

(a) *Goal.* Commanders should establish a small unit (platoon or smaller) foot movement capability of 27 kilometers per day cross-country or 54 kilometers per day on roads. Fourteen hours travel time is an acceptable goal for personnel with an individual load capability (including personal equipment) of 60 pounds. Similar goals are established for each geographical area.

(b) *Progressive training.* This training is started at the 20 kilometer road distance with a load of 15 to 20 pounds. Distances and loads are progressively increased over a 90-day period or longer. A 5 kilometer distance is covered every 40 minutes on roads and firm open ground. Progressive marches are combined. day and night. road and cross-country. Final marches are conducted at night as part of a tactical training exercise.

(*c*) *Commander's considerations.*

1. The first marches may cause 10 to 50 percent of the personnel to acquire blisters due to the rapid rate. These will occur on the heel, balls of the feet, and where the toe cap joins the sole leather of the boot. The commander should—

(a) Insure individual fit of boots;

(b) Insure that heavy socks are worn and changed regularly, especially in wet weather; and

(c) Inform personnel that blistered areas will eventually numb through continued walking and do not provide a valid reason for discontinuing a march.

2. Several preventive measures will help reduce the number of blisters:

(a) Insure proper fit of boots;

(b) Put adhesive tape on irritated foot areas as they occur;

(c) Secure the socks to boot tops to prevent slippage.

3. Commanders who execute marches on hard paved roads or other hard surfaces will find some Soldiers complaining of pain in the knee joint, hip joint, foot arch, and the small bones in the feet. The older men are normally the most affected and, in some instances, can be rendered unfit for further training. Constant bone shock from walking on hard surfaces causes these ailments. Cushion sole shoes will reduce this shock. The commander should take deliberate steps to avoid hard surfaces. Squad leaders are assigned the responsibility of keeping the men off paved surfaces when paralleling roads, especially at night. It is a natural tendency for tired men to seek the easiest walking surface.

(*d*) *Footwear.* The commander may consider the use of other than issue types of multi-purpose footwear.

(5) *Vehicle mobility training.*

(*a*) The commander is faced with several problems when the extensive use of wheeled or tracked vehicles are required for the accomplishment of his mission. Each man must be a qualified operator and be better trained than normal operational duties require. Training takes place over the most difficult terrain available in all types of weather and visibility. Personnel are taught to fire organic and personal weapons from the vehicle. At the end of the training phase each man must be able to load and move the vehicle in minutes as well as change a wheel in 2 minutes. When operating with tracked vehicles. it must be remembered that a well-trained crew will require at least 1 hour to replace a block of track under field conditions. Actions against ambushes are taught. The men as a group learn to drive at high speeds while communicating and maintaining control at the same time.

(*b*) The commander must consider logistics when selecting his operation vehicles. Once behind enemy lines, maintenance and resupply are difficult. Lightweight vehicles such as ¼-ton trucks offer good mobility in most areas of operation. They are easy to conceal, less noisy, easier to repair, do not require special mechanics, and require less POL per driven mile.

(6) *Weapons training.* The commander should, time permitting. train every man in the use of all friendly and enemy weapons. This training should be done in a practical manner. Emphasis on weight, cyclic rate of fire, and other academic features can be overlooked. Teach the basic assembly and disassembly and then let the men shoot the weapon. Each Soldier will easily learn the most effective method of operating the weapon if the commander has used a good personnel selection system.

52. Weapons Selection

a. The extended operational unit commander is normally given wide latitude in his selection of weapons. His training phase and experimentation under different terrain and type objectives will indicate the best weapons for each operation.

b. Special unit commanders in the past have designed new weapons such as combination incendiary and explosive bombs and rocket grappling hooks. Technical assistance in devising such material may be obtained through liaison with appropriate technical services.

53. Equipment and Clothing

a. Equipment and clothing used on extended operations will vary with the area of operation and availability of the items. The commander should realize that he must plan in detail for the necessary equipment for each operation to include what equipment and clothing will be common to all members of the unit. A checklist and inspections are essential. For instance, in jungle operations the wearing of undergarments can be the cause of severe skin infection. Failure to consider a small matter during the planning phase can prove deterimental to the accomplishment of the mission. Omission of individual cleaning equipment for weapons is another example. The following paragraphs contain sample equipment checklists common to extended operations. They are not all-inclusive but will provide some assistance to the commander. He will determine *those items which each individual will carry and those carried by the section, squad, or platoon.* Some items listed may prove valuable only in the event evasion or survival techniques are necessary. This is normally a consideration when planning extended operations.

b. Equipment for extended operations are—

(1) Binoculars.

(2) Camouflage sticks—use of field expedients is normally preferred.

(3) Canteen.

(4) Can opener.

(5) Candle.

(6) Compass—if possible, per each buddy team or individual.

(7) Crimpers—also may serve as field expedient wire cutters.

(8) Cup, canteen.

(9) Entrenching tool—carried for digging latrines, refuse pits, and improving water points at the patrol base; also used for preparing demolition implacements.

(10) Flashlight with batteries.

(11) First aid packet.

(12) Hammock—this is particularly useful when it is necessary to form in patrol bases, in swamp areas, or jungle but can be used anywhere. Parachute maintenance companies may provide an expedient hammock.

(13) Iodine water purification tablets.

(14) Individual survival kit—razor blades, twine, fishhooks, needles, etc.

(15) Kit, first aid (par. 56).

(16) Knives, pocket and sheath.

(17) Machete.

(18) Metascope.

(19) Matches, waterproof.

(20) Mirror, steel, small.

(21) Maps—this item disintegrates very quickly when carried on patrol. Some method of protection is required. Painting the surface of the map with a mixture of clear shellac and gum is effective. This not only protects the map, but enables the use of grease pencils on the surface.

(22) Oil, thong, gun patches.

(23) Mattress, air—useful for its intended purpose and navigating across streams and waterways.

(24) Packboard or rucksack—extended patrols often require the exclusive use of these items in lieu of the web carrying equipment.

(25) Pencil and paper.

(26) Pocket altimeter—of particular use when operating in mountainous country. Used to check heights against the map.

(27) Mess and cooking gear.

(28) Rope, climbing, toggle, 12-foot utility and snap links.

(29) Litters—improvised. A strip of canvas approximately two meters long by one meter wide with seams down the edges through which poles may be passed. It can also be used for sleeping in patrol bases.

(30) Sleeping bag, lightweight.

(31) Tape, luminous.

(32) Watch, wrist.

(33) Whistle.

(34) Web, equipment.

(35) Wire cutters.

(36) Whetstone.

c. Clothing for extended operations include—

(1) Poncho.

(2) Spare clothing, especially socks, should be considered in extended operations in order that troops may sleep dry. It is normal for wet, dirty clothes to be put on again before leaving the patrol base unless an opportunity to wash and dry out clothes has occurred.

54. Clandestine Assembly Areas and Patrol Bases

a. Clandestine Assembly Areas. These areas may be used as temporary patrol bases (see FM 21–75). They are usually bordered by difficult terrain such as swamps, steep slopes. etc.

(1) *Characteristics of a clandestine assembly area.*

(*a*) Good cover and concealment within the area.

(*b*) Adequate routes of access.

(*c*) Away from routes that offer natural lines of drift for aggressor forces.

(*d*) Sufficiently close to the area of primary operations to permit maximum use of time and facilities within the area

(*e*) Can be evacuated quickly in the event of detection.

(*f*) Has an accessible ALTERNATE AREA.

(2) *Requirements for reconnaissance.* Upon arrival at a clandestine assembly area the site must be reconnoitered to insure the following:

(*a*) The area is free of enemy forces.

(*b*) Sufficient natural cover and concealment is afforded

(*c*) The area is easy to defend.

(*d*) Early warning systems can be established.

(*e*) Provides covered routes to other assembly areas.

(*f*) An adequate water supply if site is to be occupied for an extended period of time.

(3) *Occupation of clandestine assembly area.* Upon movement into such an area the unit should—

(*a*) Establish an early warning system.

(*b*) Establish security within the site.

(*c*) Select a command post and sectors of responsibility.

(*d*) Determine work priorities.

1. Prepare defensive areas and individual sectors of responsibility.

2. Care and cleaning of weapons.

3. Distribution of ammunition and equipment.

4. Individual requirements.

5. Personal hygiene.

6. Feeding.

7. Rest.

(*e*) All items discarded will be buried and the area camouflaged.

(*f*) Prepare alternate plans for emergency use to include movement to an alternate site, selection of alternate routes and assembly areas.

(*g*) Enforce discipline relative to movement, noise, light, and fires.

b. Patrol Bases. These bases are frequently occupied for an extended period of time. They are normally removed from the vicinity of the enemy area of operations and are situated some distance from habitation. They are normally capable of being resupplied by either airdrop or special resupply patrols.

(1) *Guerrilla base.* The extended patrol may conduct joint operations with guerrilla or special force units or both. Their bases are a refinement of the patrol base and for planning purposes bear definition. The trizonal security system (Zone A, B, and C) may be used to define the three general types of guerrilla bases.

(*a*) *Zone A.* Zone A is the main base. It is secured by a regular guard system, but its safety largely depends on advance warnings from agents in Zone C, or patrol observers in Zone B. If enemy action threatens, the guerrillas move to another location prior to the arrival of enemy forces.

(*b*) *Zone B.* Zone B lies well beyond the populated Zone C in territory not well controlled by the enemy; the guerrilla force can operate overtly in this area. It is usually terrain which limits, restricts, or impedes movement. The warning system depends upon stationing observers to watch for enemy movements in the area.

(*c*) *Zone C.* Zone C, the farthest from the main base area, is usually well populated and is located inside enemy controlled territory. Enemy security forces, police, and/or military units exercise relatively effective control; the populace may be predominantly hostile to the guerrillas. At the same time, there are excellent lines of communication by which clandestine agents are able to warn the guerrillas quickly of enemy activity. This area is known as the clandestine zone and the functions of the warning system are the responsibility of the underground.

(2) *Patrol base.* The patrol base depends largely on secrecy for its security; it may be necessary to have a cover plan which will draw enemy attention from the base. These cover plans become part of the patrolling mission against the enemy. Deception should be used when possible. Some suggestions are as follows:

(*a*) When the terrain is suitable for night movement, the approach march is made during darkness.

(*b*) Centers of population should be avoided during the approach march.

(*c*) Upon occasion it may be necessary to detain local inhabitants who have observed patrols during the approach march.

(*d*) Discretion must be used in selecting base sites in order to limit the possibility of enemy detection.

(*e*) Fires by day should be smokeless. No fires should be permitted at night unless there is extreme danger of cold weather injury; then, fires should be covered.

(*f*) No more than one trail should lead into a base. The trail should be concealed, guarded, and/or mined at points easily identifiable to the patrol members. Extreme caution and commonsense are used in this latter operation.

(*g*) During movement to, and establishment of, the patrol base, noise and march discipline are emphasized.

(3) *General.* The following is true of all patrol bases:

(*a*) *Location.* Prior planning and study of map and aerial photographs will indicate possible locations. The following are factors to be considered in locating a base.

1. The MISSION of the patrol.

2. Secrecy and security of the desired base location.

3. The ability to establish required communications. Radio communication is improved when sets are sited on high ground.

4. The need for air resupply. When air resupply is anticipated, it is desirable to have a convenient drop zone.

These should not be so close to the patrol base as to endanger its security. Different sites should be considered in the event additional resupply is necessary. Air resupply should take place at night when possible. Upon receipt of the delivered supplies, movement of the friendly patrol will be slowed by heavy or bulky items. Care is exercised to insure only essential equipment is air dropped. Boobytrapping the exit trail from the drop zone will slow pursuit by the enemy if the drop has been detected.

5. Suitability of the area. Flat and dry ground with good drainage is best. Security is considered before comfort. The use of hammocks or stick platforms may facilitate comfort in otherwise poor areas.

6. Proximity to water.

(*b*) *Layout* (fig. 21).

1. The unit should establish an SOP for laying out a clandestine assembly area or patrol base. In this way it becomes routine for members of a patrol to establish such control areas. The patrol leader indicates the

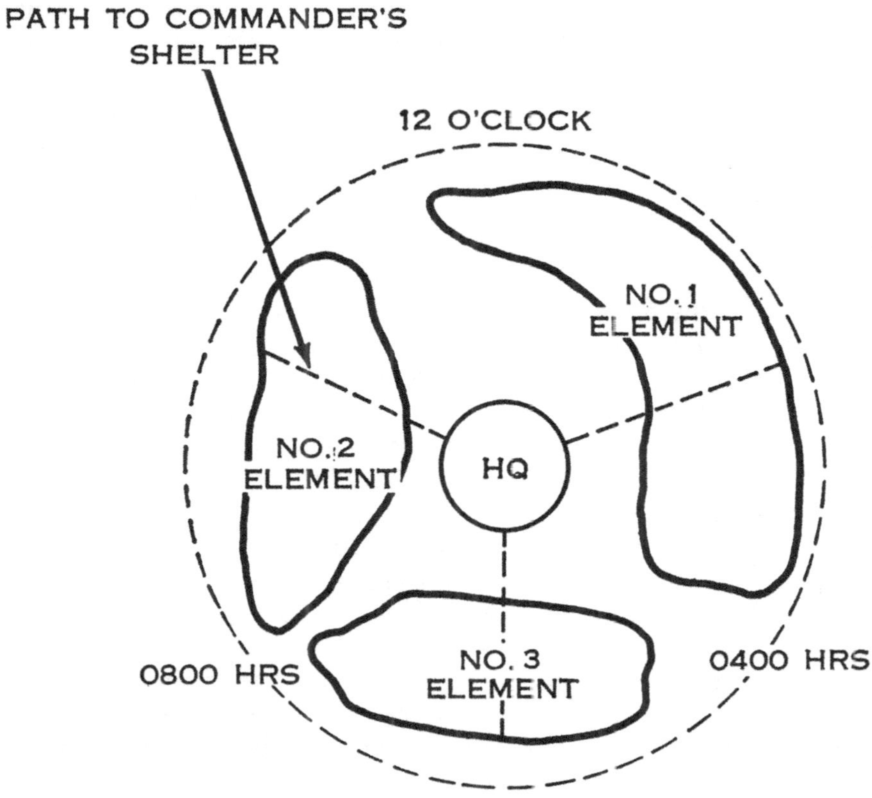

Figure 21. A suggested layout of a three-section patrol base.

center of the base and points out 12 o'clock. The men then adopt temporary positions in assigned areas which are checked by the element leaders. Necessary changes are made quickly and held to a minimum.

2. The above procedure enables each man to know his own and his neighbor's area of responsibility. The triangular concept is well suited for the organization of a patrol base. After the patrol leader indicates 12 o'clock, number one element moves up to take position between 12 o'clock and 0400. Number two element moves between 12 o'clock and 0800. Number three element moves between 0400 and 0800. Members of each element are positioned in relation to the terrain as well as to capitalize on organic weapons. As far as possible, elements are standardized.

(*c*) *Establishment.* Procedures for establishing a base are as follows:

1. Upon arrival at the selected site, local security is immediately placed out. A reconnaissance element is dispatched to reconnoiter the site and the area within hearing distance. If the reconnaissance element is broken into teams, care is exercised to preclude chance clashes between friendly elements. The remainder of the patrol takes up defensive positions within the base until the reconnaissance element returns. Upon return of the reconnaissance element, the precedure outlined in (*b*)*2* above is accomplished.

2. The patrol leader makes required adjustments.

3. Positions are prepared and personnel are informed of the approximate period the base will be occupied along with the following administrative information:

(a) Sentries required for security of the patrol base.
(b) Local security in the vicinity of the patrol base.
(c) Passwords, checks, and conduct of the alert system.
(d) Maintenance of weapons and equipment.
(e) Water and sanitation procedures.
(f) Use of cooking fires and smoking.
(g) Waste disposal and refuse pits.
(h) Other administrative requirements.

(*d*) *Security.* The base is located and protected to prevent discovery or surprise attack from the enemy. When the majority of the patrol is operating away from the base, a security guard maintains observation over the general area. This guard insures the patrol base is not discovered by the enemy. Its actions are passive. It possesses a mission of warning returning patrols in the event the enemy discovers the camp. Equipment vital to the mission is concealed away from the general vicinity of the base to prevent its loss in the event location of the patrol base is compromised.

1. Sentries are required during the day, particularly on routes leading into or by the patrol base. They are posted at the farthest point at which noises from the base can be heard. Personnel are posted in pairs when possible. Avoid posting new or inexperienced men singly.

2. Prior to darkness, sentries are drawn in toward the perimeter. After darkness, no one is allowed out of the base without the commander's authority. Sentries alert key leaders when emergencies arise. This prevents confusion while maintaining silence.

3. Local movement is carefully controlled by the commander to keep tracks to a minimum within the patrol base.

4. Every man remains armed at all times. Men move in pairs when possible. Strict discipline is required to insure that this is observed.

(*e*) *Alert procedures.* Upon indication of enemy or unusual activity within the area, the following procedures will be carried out:

1. Sentries alert key leaders.

2. Leaders alert all subordinates.

3. All personnel move to assigned defense areas and await further instruction. Normally the patrol base is not defended; an evacuation plan is prepared. However, a limited defense will take place when necessary to insure retention of equipment vital to the mission.

4. Prior to elements departing on missions, they will receive locations of rendezvous points which will be used in the event the location of the patrol base is compromised. Predetermined signals are used to keep elements from returning to the compromised location.

5. Upon movement to a secondary patrol base (alternate patrol base), increased security measures will be necessary since the enemy is aware of the patrol's operations in this vicinity.

Note. Silence is of utmost importance in carrying out the above procedures.

(*f*) *Supervision.*

1. All personnel must be thoroughly briefed prior to the start of the operation.

2. The alert plan must be known by all personnel as soon as possible. Frequent checks are required to insure complete familiarity.

3. Continuous security checks are required.

4. Thorough inspection of each position within the patrol base is accomplished as soon as possible after establishment of the base.

5. Inspections are conducted each evening. This includes all-round security of positions, equipment, weapons, and knowledge of personnel. In the event patrols are to be dispatched during the night or early morning, checks are conducted during this inspection for service ability and maintenance of equipment to be carried.

6. Element leaders check personnel prior to their leaving the patrol base; the leader of the patrol base spot checks patrols.

(*g*) *Summary.*

1. A patrol base is located in an area from which its elements can carry out assigned tasks. The exact location is decided by the requirements of the mission; maximum comfort is gained in consonance with security.

2. The extent to which a patrol base is developed depends upon the period of occupation.

3. Security and comfort are achieved through strict discipline and able leadership.

4. Some considerations used in establishing a patrol base are as follows:

(a) A well-planned alarm and evacuation system is known to and practiced by everyone.

(b) Adequate security is maintained both actively and passively.

(c) Duties are shared to the maximum.

(d) Strict rules of hygiene and sanitation and water discipline are enforced.

(e) Cooking and food preparation are in keeping with the danger of detection by the enemy.

55. Communications

a. Radio Communications. The extended operations commander will often be faced with establishing a dependable long-range radio communications system. For operation up to 50 kilometers, present tactical FM radio sets can be used. It may be necessary to use the jungle antenna RC–292 or a field expedient type antenna (FM 24–18). Using this type of antenna, the rated range of radio set AN/PRC–8, 9, 10, and RT 66, 67, and 68 will be doubled or even tripled depending on the terrain. When radio communication must be established at ranges greater than 50 kilometers, it will be necessary to use medium and low frequency radio sets such as radio set AN/GRC–87 or RS–1. The radio set AN/GRC–87 has both voice and CW emission, the RS–1 is CW only. CW is much more dependable over long ranges than voice, but the operator must have special training (MOS 050 or 051) when using CW. The commander must solve his communication problems before his unit is completely operational. If radio communication must be established at ranges greater than 50 kilometers, he should take the following into consideration:

(1) Obtain qualified CW operators (MOS 050 or 051). It is desirable that these operators have a repair background.

(2) Select the most suitable radio set for the type of operation. The terrain, range, and type emission desired must be considered in selecting the radio set.

(3) Incorporate communications into every phase of his training to insure that he has various workable communications plans to fit every conceivable type combat mission.

b. Personnel Selection. The Ranger unit (reconnaissance) will require one radio operator (CW low speed or FM voice) per three-man team to accomplish medium-range surveillance patrol missions. Since trained radio operators are critical in a combat zone, the commander should solicit volunteers from his unit to be trained as radio operators. These radio operators do not have to be school trained, but can be trained in the unit, using very simple equipment such as tape recorder RD–94 and code training set AN/GSC–T1. The commander need not be concerned with the fact that his personnel may not have a high code aptitude score as these operators need not be trained to transmit and receive international Morse code in excess of about 10 words per minte.

c. Selection of Radio Sets.

(1) For short-range surveillance patrol missions, the commander would select one of the standard tactical FM radio sets. For medium- and long-range patrols, he must use either the AN/GRC–87 or RS–1. The AN/GRC–87 may provide communication up to 1,500 kilometers (using a good antenna system), but at this range special consideration must be given to frequencies used, atmospheric conditions, and the time of year. Detailed information on selecting frequencies for long-range communication can be found in TM 11–666 and radio propagation charts which can be procured from United States Army Signal Radio Propagation Agency, Fort Monmouth, N.J. These charts are published monthly and must be requested for the particular area of operations. The commander may desire to contact a Signal Corps unit operating in his area for information and guidance on the selection of radio sets, frequencies, and antennas.

(2) FM radio sets may be used for interpatrol communications, air-to-ground communications, and air-to-ground relay. The receiver-transmitter RT–66, RT–67, and RT–68 have a radio navigational beacon capability. If air-to-ground relay is necessary or desired, the commander should contact his supporting headquarters and have them establish an aircraft relay system. This may require more than one aircraft, depending on location and operating range. An aircraft at an altitude of 500 meters can relay messages from the patrol up to 100 miles.

d. Antennas.

(1) Regardless of how good or what type of radio sets the commander may select, the "heart" of any radio is its antenna.

(2) The whip antenna was designed for mobile and portable operation, but is ineffective for dependable long-range communications for both AM and FM radio sets. Effective antenna systems can be constructed in the field using available material for both AM and FM radio sets.

(*a*) *Field expedient directional antennas for FM radio sets.* The wave antennas and the vertical half rhombic antenna are the two field expedient directional antennas which can be used with the frequency modulated radio sets. These antennas can be easily constructed, using field wire (WD–1) and poles or existing trees as supports. These antennas are directional and will transmit and receive in the direction of the terminated end. By removing the terminated resistor, the antenna will radiate and receive in both directions along the axis of the antenna. These antennas will normally double or triple the rated operating range of the FM sets.

(Note section IV, FM 24–18 for illustration and construction details.) Other types of antennas and construction details are outlined in section IV, FM 24–18, 1 July 1958.

(*b*) *Field expedient directional antennas for AM radio sets.* There are many different types of antennas that can be used with AM radio sets for long-range communications. TM 11–666 and FM 24–18 can be used by the commander to select the best type antenna for the particular situation. A type antenna that may be employed under most operating conditions is the center fed half wave or doublet antenna. This antenna, when used with the low power AM sets such as the RS–1 and AN/GRC–87, can be constructed with available material (WD–1 field wire); the insulators for this antenna can be made from pieces of dry wood and need not be anything elaborate. Notice in figure 22 the antenna proper is divided into two parts, each a ¼-wave long; the length of these two sections is very critical and is determined by the frequency that you intend to use. To determine the length of the two ¼-wave sections, use the following formula:

$$\text{Length in feet} = \frac{234}{\text{FREQUENCY IN MCS}}$$

(Or 234 divided by the frequency in megacycles.)
Example: operating frequency—2.0 mcs.

```
    117
2.0/234  each ¼=wave section should be 117 feet long.
    2
    --
    03
     2
    --
     14
     14
     --
      0
```

IMPORTANT. THIS ANTENNA SHOULD BE AS HIGH AS POSSIBLE FROM THE GROUND. 25 TO 60 FEET SHOULD BE SUFFICIENT. IN AN EMERGENCY, SATISFACTORY RESULTS COULD BE OBTAINED WITH ANTENNA LOWER THAN 25 FEET, BUT THE HIGHER THE BETTER. THIS IS A DIRECTIONAL TYPE ANTENNA AND THE SIDE OF THE ANTENNA SHOULD BE TOWARD YOUR RECEIVING STATION. The length of the transmission lines is also important, and the transmission lines should be adjusted to obtain the best results. This is accomplished by installing transmission lines that are longer than necessary, and then by shortening them two feet at a time until best results are obtained. When operating at ranges greater than 100 kilometers, changing atmospheric conditions will greatly affect the satisfactory operation of this type communications systems. The commander, aware of this fact, will be unable to understand why communications fail when the radio operated previously. The signal officer or communication officer using the radio propagation charts mentioned in *c* above should determine which operating frequencies should be used at different times of the day and a schedule for radio contact should be established before departing on a patrol.

Figure 22. Simple sky-wave antenna.

The receiving headquarters should monitor all the operating frequencies at the prearranged time.

(c) *Siting of radio sets, AM and FM.* The siting of a tactical radio set and its associated antennas usually is a compromise between technical requirements and tactical consideration of cover and concealment. A radio station should be placed in a position which will assure communication with all stations with which it must operate. The transmitter and receiver will have a greater range if the antenna is located in a position which is high and clear of hills, buildings, cliffs, densely wooded areas, and other obstructions. Depressions, valleys, and low places are poor for radio transmission and reception because the surrounding high terrain tends to impede the signal. Weak signals may be expected if the radio set is operated close to steel bridges, underpasses, or near power lines or power circuits. The antenna should not be allowed to touch trees, brush, or camouflage material.

e. Conclusions.

(1) It is essential that intelligent men are selected for radio operators. Success of the patrol's communications is dependent on the training of radio operators, on the preparation and checking of radio sets and power sources, and on the

intelligent selection of radio sites by commanders. Once radio operators are committed to an extended operation, they will be without technical assistance.

(2) Radio communication on extended operations, although often difficult, is rarely impossible. With constant training an operator soon learns to have confidence in his set and his own ability to establish communication under the conditions and ranges involved.

(3) The commander must realize that although the higher headquarters signal officer can offer some assistance, it is the commander who must be thoroughly familiar with communication problems and their solutions.

56. Medical Considerations

During extended patrol operations behind enemy lines, the commander often suffers casualties. If no medical troops trained and conditioned for patrol activities are available, primary reliance must be placed on first aid measures performed by combat members of the patrol. When forming his unit, the commander should consider selecting volunteers who have a medical interest. In any event, it is essential that every man should understand not only the basic principles of hygiene, sanitation, and first aid, as expressed in FMs 21–10 and 21–11, but also more advanced techniques of emergency care outlined in TC 8–1. This in particular, applies to senior and junior leaders who are responsible for the health of their men. Many a Soldier has been saved from death or permanent disability because immediate emergency care was rendered; many have died because their comrades lacked the knowledge or confidence to apply proper first aid measures. The extended unit commander should provide suitable time for the unit surgeon or his representative to brief his personnel concerning health hazards of a particular area and medical techniques which may be needed. The unit surgeon may be able to provide further guidance during the operation by use of radio or other communication means.

a. Disease. Communicable diseases cause more casualties during wartime than do wounds and injuries. Communicable diseases begin with some of the same general symptoms such as headache, loss of appetite, weakness, chilly sensations, nausea, and fever. Later other symptoms usually appear which are related more specifically to the disease process present. As a rule, diseases caused by bacteria develop approximately 4 days after exposure and those caused by virus, 14 days after exposure, but there are wide variations within these general guides. Diseases which develop during a prolonged patrol operation may have been contracted either during or prior to the beginning of the operation.

(1) *Prevention of disease.* Significant manpower savings often are possible by following the basic preventive techniques for avoiding disease, or controlling an outbreak within the unit.

(*a*) *Respiratory disease.* Where individuals mingle, respiratory disease germs from coughing, sneezing, or even natural breathing may remain alive in droplets of moisture on dust particles in the air. These then can enter the nostrils or mouths of new victims. Typical respiratory diseases are head colds, sore throat, sinus infection, bronchitis, pneumonia, and influenza. Preventive measures consist of dispersing individuals as much as possible, particularly during hours of sleeping; by adequate ventilation of the quarters (bunkers, buildings, caves, etc.); and when possible by constructing a "sneeze sheet", a screen made from a shelter half or poncho, erected between bed rolls.

(*b*) *Intestinal disease.* Under field conditions the greatest single disease risk is intestinal disease, transmitted mechanically (food, feces, fluids, fingers, and flies) from the sewage of infected individuals to the mouths of new victims. A common source is drinking water from streams and rivers which are contaminated by latrines or night soil. Typical examples of intestinal disease, which are usually associated with vomiting, diarrhea, and fever are dysentery, cholera, and typhoid fever. Preventive measures consist of purifying all drinking water with iodine tablets or chlorine powder, eating only issued individual rations or thoroughly cooked foods, and washing the hands before eating and on leaving the latrine.

(*c*) *Insectborne disease.* The bites of various insects can introduce certain diseases through the skin. Typical examples are from mosquitoes (malaria, yellow fever, dengue), lice (typhus, trench fever), ticks (spotted fever, Q fever), mites (scabies), chiggers (scrub typhus), fleas (plague, murine typhus), and sandflies (sandfly fever). Preventive measures consist of using insect repellent and insecticides to keep biting insects away. In the case of malaria, which is the most prevalent disease in the world, the best effective single measure is taking a suppressive drug, such as chloroquine, once weekly to prevent the onset of malaria symptoms.

(2) *Treatment of disease.* If it is tactically possible, disease casualties should be allowed to rest where they can remain relatively comfortable and can obtain food and water. If recognized, a specific disease should be treated according to

medical guidance available. Otherwise, the general treatment of disease is symptomatic. For example, if the individual has diarrhea, he should be given some preparation to decrease the number of movements; if he is feverish, he should be given medicine such as aspirin to lower the fever artificially; if he has headache or other pain, he should be given pain medication; if he is shaking with chills, he should be bundled up with extra clothing, covered with a poncho or shelter half, and given warm drinks; if he is sweating profusely, he should be dried off frequently and provided with a change of clothing or covering so that he will not later be chilled by the wet clothes.

b. Cold Weather Injuries.

(1) World War II produced 73,740 cases of trench foot and frostbite in the European Theatre alone among United States Forces. These injuries could be attributed to weather, clothing, type of action involved, and lack of preventive measures taken on the part of the individuals and leaders. Cold weather training is a prerequisite for extended operations in cold-wet or cold-dry weather.

(2) Frostbite is caused by exposure of the body to the cold. The fingers, ears, toes, and face are most likely to be affected. Frostbite is not painful at first. The first indication of its presence is a numbness or whitish patch on the skin.

(3) Frostbite can be prevented; however, when it happens it requires medical treatment. Proper dress to include head and footgear is the first consideration. Getting wet from an outside source should be avoided. Care is taken to prevent sweating by ventilating or removing clothing by layers. The hands and feet are exercised during inactive periods or halts. Each man watches his buddy's face for white patches on exposed skin. These frozen spots are thawed by placing a warm hand over the surface. *Never attempt to thaw a frozen spot by rubbing it with snow.* Make faces or grimace to exercise your face.

(4) Trenchfoot is caused by prolonged exposure of the feet to cold and moisture and is usually associated with lack of movement and constriction of the limbs by shoes or clothing. It is more frequent in cold, wet climates than in cold, dry weather.

(5) It some times takes 36 hours for trenchfoot to develop. Like frostbite, little or no pain is noticed at first, but the feet feel numb or become hard to control. In later stages the feet swell, become red, and begin to hurt. In the final stage of trenchfoot, the flesh dies and the foot may have to be amputated.

(6) Be on guard constantly for trenchfoot. Keep the feet dry. If they get wet, warm them with your hands, apply foot powder, and put on dry socks. When boots or socks are wet or when the weather is very cold, never keep your feet still for long. Jump up and down or double time a few steps back and forth. If you cannot do this, use foxhole exercises, such as flexing your muscles and moving your arms and legs. Flex and wiggle your toes in your boots. Keep entire body warm and dry, avoid wearing clothing that is too tight.

c. Principles in the Care of Wounds. The basic essentials of first aid can be summarized as stop the bleeding; be sure airway is open; protect the wound with a dressing; apply a padded splint to any broken bones, to include the joint above and the joint below; and position the casualty to the best advantage. It is always important to handle the casualty gently, and to provide all possible comfort, encouragement, and reassurance. In any wound there are several possible complications which may require special consideration.

(1) *Shock.* Shock ("traumatic shock" or "surgical shock") is a weakening of the body associated with decreased blood circulation. (This condition should not be confused with "emotional shock," "electrical shock," or "shell shock" which are names associated with other conditions.) In general the causes, symptoms, and treatment for shock can be understood best by considering it a condition in which there is not enough blood to fill the blood vessels.

(*a*) Too little blood in the blood vessels can result from bleeding, or in the case of burns and some other injuries, from loss of the fluid portion of the blood through the vessel wall into the damaged flesh. Or with the usual volume of blood, there may be an enlargement of certain blood vessels as a result of severe pain, with the result that there is not enough blood to fill up the total volume of the blood vessels. Therefore, bleeding, burns, and pain are three common causes of shock.

(*b*) The symptoms and signs of shock can be related to this basic mechanism: a weak pulse at the wrist or in front of the ear (with insufficient blood there is reduced blood pressure); a fast pulse (the heart tries to maintain the blood pressure and the flow of blood by beating faster); paleness and cold skin (small vessels in the skin do not fill with blood); thirst (the body borrows water from the flesh to make up some of the shortage of blood); and weakness, faintness, or lightheadedness, often progressing to restlessness or to mental sluggishness and confusion (there is insufficient circulation to the brain to keep it alert). Shock can become progressively worse as a form of dying and every possible effort should be made to reverse the process. In the field the best index of degree of shock is judging the force of the pulse at frequent intervals, and counting the number of beats per minute; if the pulse becomes faster and weaker, the casualty's condition is worse and if the pulse becomes slower and stronger he is improved. Sometimes the pulse is too weak to be detected at the wrist or in front of the ear, and the rate of the heart beat can be determined only by placing the ear on the casualty's chest.

(*c*) Under field conditions the following measures are valuable in preventing or treating shock: stop the bleeding and relieve severe pain (to remove the contributing causes); position the casualty with his legs higher than his head and chest (to mobilize more of the available blood to the vital areas of the body); encourage casualty to drink water (to provide temporary fluid to the body); and cover the casualty with clothing or a poncho (to retain body heat since there is not enough circulation to keep the entire body at desirable temperature). Additional heat, in the form of warm stones or canteens, should *not* be applied to the casualty since they cause a local dilation of blood vessels which could contribute to the overall condition. As soon as the casualty can be gotten to medical personnel, ultimate treatment consists of transfusions of blood or of some blood substitute such as dextran, plasma, glucose solution, or saline solution.

(2) *Hemorrhage.* Loss of blood is a major cause of death in war, either by direct bleeding to death or as a cause of severe shock. In approximately 95 percent of all wounds, bleeding can be controlled by a snug dressing. However, when it is decided that only a tourniquet will control bleeding from an arm or leg, the tourniquet should be applied at once and it should be tight enough to stop all blood flowing to the arm or leg. It should be left in place until medical advice is available. The length of time that the flesh below the tourniquet may remain alive depends on several factors, the most important being temperature; in cold weather the safety period is longer than in hot weather. Statistics indicate the risk to casualty caused by additional bleeding from a loosened tourniquet is greater than that caused by gangrene of an extremity.

(3) *Infection.* In any open wound there is the hazard of infection which becomes apparent within a few days. In usual combat the casualty reaches medical facilities before infec-

tion appears, but in extended patrol operations he may still be with the combat commander when serious infection develops.

(*a*) *Symptoms.* The usual signs and symptoms of infection are pain, redness, swelling, and local warmth. Severe infections often cause generalized sickness in the form of fever and marked weakness. Often pus forms in the infected wound and if allowed to collect it may penetrate, carrying infected material to new areas.

(*b*) *Treatment.* Antibiotic medicines assist the body in its defenses against infection. Each antibiotic has certain harmful features which in peacetime are evaluated carefully before the drug is prescribed, but under wartime conditions antibiotics may be used more routinely as a calculated risk. Where pus pockets can be seen it is usually helpful to make one or more cuts in them with a sterilized knife, so that adequate drainage can take place. Often the local application of mild heat several times a day, as with warm canteens or hot wet cloths, improves local circulation and encourages an infection to localize toward the surface.

(*c*) *Cleanliness.* Normal washing or soaks in warm water are not harmful to an infected wound, but every effort should be made to avoid introducing further contamination. No special effort should be made to kill maggots which may infest a wound, since they tend to remove dead tissue and from that standpoint are beneficial to the casualty.

(*d*) "*Blood poisoning.*" An occasional complication of an infection is the appearance of lymphangitis (lymph poisoning, often called "blood poisoning"), which may be seen as red streaks progressing from the site of the infection toward the heart. Usually there is associated soreness and swollen lymph nodes along the path. For example, at the front of the elbow or in the armpit, with an infection at the hand; or at the back of the knee or in the groin with an infection at the foot. Besides treatment of the original infection, hot applications, elevation, and/or loose splinting of the affected arm or leg are helpful.

(4) *Severe pain.* Usually such measures as protecting a wound with a dressing and splinting any fracture or severe wound provide sufficient relief from pain; but occasionally there will be cases of severe pain which can be relieved only by morphine. Morphine syrettes may be available to nonmedical troops outside CONUS for the emergency treatment of severe pain.

(*a*) *Use of morphine.* Morphine syrettes can be injected into buttocks, the thigh, the upper arm, or the deltoid or pectoral areas of the chest. The usual dosage is one syrette, but the policy of its use, its dosage, and the method of marking the casualty to whom it has been given will be specified by the commander on the advice of the unit surgeon.

(*b*) *Morphine actions.* The various actions of morphine should be understood by all individuals who use it. These are relief from pain; slowing of breathing; slowing or stopping certain nerve reflex actions; and slowing or stopping the usual intestinal movements. It is appropriate for the relief of severe pain or for wounds into the abdominal cavity.

(*c*) *Harmful use of morphine.* In general, morphine should not be used for wounds above the belt. In head wounds (in which there is usually little pain) morphine tends to alter various nerve reflex actions and prevents thorough evaluation of the casualty's condition. Morphine should not be used in cases of unconsciousness since there is no evidence of pain. In any wound of the neck or chest where the casualty has difficulty breathing, morphine should not be given since it depresses breathing further. It is not suitable as a sedative for ordinary apprehension.

(*d*) *Morphine overdosage.* In warfare there is a risk of overdosage. This occurs when the slowed circulation of a casualty in shock fails to absorb the morphine from the site of injection and additional morphine is given later. As the casualty recovers slightly from the shock, his circulation may absorb two or more doses simultaneously, with resulting overdosage. Nonmedical personel should not administer a second injection of morphine without at least radio contact with medical personnel.

(*e*) *Other drugs.* If other narcotics are available or synthetic substitutes for narcotics, their use is essentially the same. Certain medicines which tend to relax nervousness may be effective in relieving pain also, since apprehension and tension often aggravate pain. Examples of this type of drug are phenobarbital, tranquilizers, and pyribenzamine.

d. Care for Specific Types of Wounds. In addition to the basic essentials of first aid, special considerations are indicated in certain types of wounds.

(1) *Scalp lacerations.* Tearing wounds of the scalp are characterized by profuse bleeding, but usually this bleeding can be controlled by the application of a snug dressing.

(2) *Brain injury.* In any wound of the head the principal risk is damage to the brain. This may be temporary, as in the

case of "bruising" of the brain with temporary swelling, or it may be permanent, as in the case of actual destruction of brain tissue. Ordinary first aid measures are usually limited to elevating the head slightly higher than the shoulders and hips, and applying cool cloths to the head. Since unconsciousness often is present or may develop, the casualty should be positioned to avoid his choking on saliva, blood, or vomitus. Morphine should *not* be given.

(*a*) *Concussion.* Severe jarring of the head may cause injury with swelling of the brain, but within the rigid, bony box of the skull there is little room for expansion. Therefore, with any swelling the brain presses against itself, interfering with its own functioning and with local blood supply. Unconsciousness often results, until the swelling subsides or until it finds an outlet for expansion. There is no special first aid measure other than those for any brain injury.

(*b*) *Skull fracture.* With fracture of the skull there is concussion and in addition, portions of the skull may be forced in to crowd the brain even more. In the absence of a visible wound, skull fracture can be suspected when there is unconsciousness associated with unexplained bleeding from the nose or ears.

(*c*) *Open wounds of the brain.* Wounds opening into the skull should be covered with a sterile dressing in order to protect the wound and to prevent introduction of further contamination.

(3) *Eye wounds.* Principal risk of eye wounds is that of permanent blindness. A dressing should be placed over the injured eye (with the lid closed) using sufficient thickness and snugness so that there can be no blink movement. It is sometimes helpful to cover the second eye in the same way to prevent the blink reflex entirely, but on an extended patrol operation nothing should be done which might prevent a casualty from taking care of himself as much as possible. On the basis of medical consideration only, eye wound cases have high priority for early hospital care.

(4) *Jaw wounds.* In the early treatment of jaw wounds bleeding is controlled by an external dressing and the casualty is positioned in such a way that saliva, blood, and vomitus which might collect in the mouth will drain out away from the throat. Morphine may be necessary for severe pain, but otherwise it should not be used since it may cause vomiting. Later if fracture is suspected, the jaws should be bandaged together in a clenched-teeth position, taking care that this

bandaging can be quickly removed by the casualty if he should vomit. Food should be limited to only those soft items which the casualty can push through his teeth with his fingers.

(5) *Chest wounds.* The mechanism of breathing depends on a vacuum between the outside of the lung and the inside of the chest, so that as the chest cavity enlarges (by the rising of the ribs and the dropping of the diaphragm or chest floor) air is brought into the lungs through the windpipes. In the case of a "sucking chest wound" there is an artificial hole through the chest wall; as the chest cavity enlarges in the normal breathing effort, air enters and leaves the chest through the wound and very little air comes through the windpipes into the lung where it is needed to maintain life. The early risk is death from lack of air in the lungs and essential treatment is an airtight dressing over the wound(s) to prevent the entrance of air. An excellent method is to place the metal foil wrapping of the first aid dressing directly over the wound and cover this with the dressing itself. (Over a period of weeks, air in the chest cavity becomes absorbed and the original vacuum is reestablished.) Later risks are from infection and from injury of organs inside the chest. On the basis of medical considerations only, sucking chest wound cases have a high priority for early hospital care.

(6) *Treatment of blisters.* Blisters that develop on the feet should be treated by washing the blister and surrounding area with soap and water; opening the blister, sticking the lower edge with a needle or knife point sterilized by a flame; and covering the area with a band aid or adhesive plaster.

e. Medical Aid Kit. The composition of aid kits prepared for extended operations will vary with the type operation and climatic conditions. The number of each item carried must be determined by the nature of the operation and size of the unit involved. The commander should consider the following items when preparing his equipment lists. Final selection and the detailed direction for their use must be obtained from the organization surgeon or other medically qualified personnel.

Stock No.	Nomenclature	Remarks
	MEDICINES	
6505–106–0875	Ammonia inhalent, aromatic..	Break tube and inhale as mild stimulant.
6505–663–2636	Sodium chloride and sodium bicarbonate mixture, 4.5g, in 2's.	First aid kit, survival, individual.

Stock No.	Nomenclature	Remarks
	MEDICINES—Continued	
6505–680–2787	Antivenom kit, polyvalent, 1 dose.	
6505–660–1720	Dextro propoxyphene HCL caps, 32 mg, 100's.	Take 1 every 4 to 6 hours for moderately severe pain.
6510–203–1175	Copper sulfate, pad, cloth, 3's	Wet pad and cover white phosphorus burn.
6505–114–8985	Codeine sulfate, 32 mg, 100's	1 tablet for severe pain.
6505–146–2200	Sulfadiazine tablets, USP, 0.5 gm, 1000's.	Sulfa. 2 tablets 4 times a day with extra water.
6505–133–9600	Phenobarbital tablets, USP, 32 mg, 100's.	1 tablet to relax apprehension of casualty. 3 to encourage sleeping.
6505–132–3030	Opium, tincture, camphorated, USP, 1 pint.	1 bottle cap full in water every bowel movement, to relieve diarrhea.
6505–116–1750	Detergent, surgical (Phisohex)	Use as soap in washing wounds or for mild skin disease.
6505–113–9295	Chloroquin phosphate tablets, 0.5 gm, 100's.	1 tablet per man per week.
6505–105–8905	Aluminum hydroxide gel, dried, tablets.	1 to 3 tablets for upset stomach. 1 bottle per platoon.
6505–111–1200	Calomine lotion, modified, 2 oz.	For irritated skin. 2 bottles per squad.
6505–112–9010	Cascara sagrada extract tablets, NF, 100's.	1 or 2 tablets as laxative.
6505–141–8000	Sodium bicarbonate, charcoal and peppermint tablets, 1000's.	Sodamint for upset stomach.
6505–147–0300	Tar compound ointment, modified, 1 lb.	Pragmator for skin infection.
6505–147–1720	Tetracaine ophthalmic ointment, ⅛ oz, 12's.	Local anesthetic for painful eye. 1 tube per squad.
6505–153–8750	Acetylsalicylic acid tablets, 1000's.	Aspirin.
6505–153–8848	Fungicidal ointment, 1 oz, 12's	For athlete's foot and other fungus infection. 2 per man.
6505–299–8276	Oxytetracycline tablets, 0.25 gm, 100's.	Antibiotic. 1 tablet 4 times a day.
6505–161–2450	Tetanus toxoid, alum precipitated, USP, 5cc.	
6505–129–5517	Morphine injection, USP, 16 mg.	For severe pain.
6505–133–0790	Pentobartibal sodium tablets, USP, 100 mg, 500's.	1 for sleep.
6505–299–8175	Sodium sulfacetamide, ophthalmic, ointment, 30%, ⅛ oz, 12's.	For eye infections.
6505–148–9000	Tripelennamine hydrochloride tablets, USP, 50 mg, (pyribenzamine).	½ to 1 tablet every 4 to 6 hours for allergy. ½ to 1 tablet may relax apprehension in a casualty.

Stock No.	Nomenclature	Remarks
	HYGIENE AND SANITATION SUPPLIES	
6840–242–4229	Insecticide, powder, louse, 2-oz can, 10% DDT.	
6850–270–6225	Water purification powder, chlorine, 100-tube box.	
6850–264–5904	Water purification tablet, iodine, 50-tablet bottle.	
6505–126–3407	Salt, tablet, type 3, class 3, impregnated, 500's.	
6840–290–5027	Repellant, insect, 2-oz bottles	
6840–254–8770	Insecticide, aerosol, 12-oz	
6840–285–1822	Warfarin	Mix with grain or dry food as rodent poison.
6505–515–1584	Foot powder, fungicidal, 1 oz	For feet and chafing areas. 2 per man.
3740–252–3384	Mousetrap, spring, wood base, E–100.	
3740–260–1398	Rattrap, spring, w/4-way release action, wood base, E–50.	
8415–261–6630	Hat and mosquito net (head net).	1 per individual.
7210–266–9736	Insect bar, nylon netting, field type (bed net).	1 per individual.
6840–270–8172	Disinfectant chlorine food service	
	SUPPLIES FOR CARE OF WOUNDED	
7210–715–7985	Blanket, bed, wool, olive green, 66- x 90-inches.	
7210–160–0370	Blanket, bed, wool, 3¾-lb, finished weight, olive drab, 84- x 66-inches.	
6510–201–2890	Compress and bandage, field, 18 x 22 compress.	1 per squad.
6510–201–2900	Compress and bandage, field, 22 x 35 compress.	1 per squad.
6510–201–7425	Dressing, first aid, field, camouflaged, 11¾-inches square.	2 per squad.
6510–201–7430	Dressing, first aid, field, camouflaged, 7½- x 8-inches.	Large wound dressing, one per man.
6510–201–7435	Dressing, first aid, field, camouflaged, 4- x 7-inches.	Small wound dressing. One extra per man.
6510–201–1755	Bandage, muslin, 27- x 52-inches.	Cravat or triangular bandage. One per man.
6510–597–7468	Bandage, absorbent, adhesive, ¾-inches x 3-inches, 18's.	Band aids.

623989°—62——8

Stock No.	Nomenclature	Remarks
	SUPPLIES FOR CARE OF WOUNDED—Continued	
6510–203–5000	Adhesive plaster, 3-inches x 5 yards.	1 per squad.
6530–783–7205	Litter, folding, folding pole, aluminum pole.	
6530–784–3120	Litter carrying strap, 90-inches, 3.83 lb, 1.425 cubage.	
6530–597–9470	Marker, casualty, arctic.	
8465–753–3237	Sling, universal, individual, load carrying, adjustable.	
6515–680–0887	Splint, wraparound, leg, arm, back, and neck; canvas, padded.	
6515–373–2100	Splint, wire, ladder, 3½- x 31-inches.	
6545–919–7675	Aviator first aid kit, camouflaged.	
6530–919–5700	Evacuation bag, casualty, insulated.	
8105–132–9026	Bag, waterproof, w/shoulder straps, 12- x 9-inches x 1¼-inches.	
	NONSTANDARD ITEMS	
PX	Pocket knife with large and small blades.	L1 per man.
PX	Dial soap	Slightly antiseptic soap for routine use and for minor skin irritations. One bar per man.
PX	Brushless shaving cream	For minor skin abrasions and burns. One tube per man.
PX	Aerosol shaving cream	For local application to painful burns. One can per squad.
PX	Small tweezers	One per squad.
PX	String, cord, or rope	
Store	Neckerchief, scout	
PX	Sunglasses	
Store	Small lens	To locate foreign bodies slivers, etc. One per squad.
PX	Small mirror	To assist in self-aid. One per man.
Store	Canvas water bags	To keep drinking water cool; similar to lyster bag but smaller.

57. Logistics and Resupply

a. The extended patrol unit, in all probability, will be attached to a headquarters situated near supply depots in the combat zone. Because of the nature and priority of the missions, the commander should not have difficulty in obtaining logistical support. He may have trouble obtaining nonstandard items, issue items not available in the theater of operations, and resupply of items when involved in a deep penetration operation.

b. The problem of nonavailability of nonstandard items and other items not in the theater can generally be solved through the use of expedients. Rear area depots have a limited manufacturing capability. Use of local civilian facilities should not be overlooked. During World War II an extended operation under General Wingate was hampered due to the nonavailability of aerial resupply equipment. Reed baskets designed to fit inside one another were manufactured by the local natives. These provided antishock effect and were used to successfully drop sensitive items without parachutes.

c. The present doctrine on resupply by air or patrols is adequate and workable. The problem is one of communications. Each minute of radio transmission endangers the security of the patrol. To solve this security problem, the commander takes three actions.

(1) First, a brevity code is devised. Every item including shoes to radio tubes is given a code symbol. Each man in the unit is also given a code symbol. Thus, if Brown needed his boots replaced, his code number would precede the code for boots. His size being known at the unit, boot sizes would not be transmitted.

(2) Certain supplies are prepacked before the operation commenced. These packages are then delivered on a prearranged schedule or on call by package number. The total supply for the operation is computed and then divided into the carrying capacity load of the patrol. A request for X bundle would be all that would be necessary to fulfill ration requirements for a certain period.

(3) Supervision of the handling of the resupply items at the friendly rear base should not be entrusted to anyone but organic personnel. The commander preceding the operation, therefore, selects one or two organic personnel to stay behind for this purpose. This is no different than assigning liaison personnel to higher headquarters as is normally done by conventional units.

d. In all foot mobile extended operations, equipment carried for the mission is a critical consideration. When resupply is impractical, food must be substituted for equipment. A partial solution exists for the well-trained Ranger unit by the use of survival or condensed rations. These rations exist and are available. The commander should test the effectiveness of these rations during his training period for future guidance as to the amount required for his particular operations.

Section VI. SMALL UNIT WATERBORNE OPERATIONS

58. General

This section outlines the considerations in planning and techniques utilized in small unit waterborne operations and river navigation.

a. Waterborne Operations. In planning a waterborne operation, the backward planning techniques must be used. The element of surprise is an essential part of the successful waterborne operation. In order to increase the opportunity of achieving surprise at the objective, the leader of a waterborne operation carefully considers the time and location of landing. If the raiding unit is prepared to encounter natural obstacles on the beach and en route to the objective, the probability of surprise at the objective is greatly enhanced. Prerequisites for the raiding unit should be proficiency in mountaineering, or cliff assault techniques, and the ability to swim. These individual capabilities coupled with the necessary special equipment make a physically difficult beach landing with obstacles between it and the objective an ideal route for a well trained unit.

b. River Navigation. River navigation techniques include the use of issue and field expedient type craft and the proper organization and utilization of personnel. See paragraph 60.

59. Planning Considerations—Waterborne Operations

a. Tactical. Of primary interest, to the commander, is the number of personnel required, and amount and type of special equipment necessary to accomplish the mission. Raid type missions normally require equipment not organic to the rifle company or platoon. Examples of these requirements are mountaineering equipment for cliff scaling, ropes for river crossings, mine detectors for breaching minefields, and Navy underwater demolitions teams (UDT) to clear the beach of obstacles and mark the beach for the landing (FM 110–115). The location of the landing site must be considered with respect to the ease with which it can be defended by hostile forces, the degree of security which the supported unit commander can employ during the landing, and the amount of cover and concealment afforded the boats while the raid is being conducted. If the landing is to be conducted on an unknown shoreline, the use of personnel from other services such as UDT men, special forces agents, or even friendly partisans should be considered to mark the exact site with lights, homing signals, or panels (fig. 23).

Figure 23. Difficult landing sites often enhance security. Shown above is a portion of the Pointe Du Hoe secured by the 2d Ranger Battalion on D-day. One of its rope ladders attached to a grapnel carried over the cliff by a rocket was still in place when photo was taken a year later.

b. Fire Support.

(1) *Naval.* If the raid is of the hit-and-run variety depending on surprise, speed, and violence of execution to accomplish the mission and withdraw from the beach, naval gunfire is of prime importance in delaying hostile forces that attempt to penetrate the beachhead. If the objective is out of range of both conventional and nuclear artillery and friendly air cannot be made available, then naval gunfire will be the only heavy support available to the raiding force.

(2) *Artillery.* Both conventional and nuclear artillery should be coordinated if the objective is within range. Should the objective be well inland, harassing rounds should be coordinated for several days prior to operation. When the operation is actually conducted, concentrations can be used, on-call, for navigational assistance to the supported unit if necessary. As a deceptive measure, artillery can be used on the objective in a destruction role to deceive the enemy as to the mission.

(3) *Tactical air support.* If the enemy does not have air superiority and the objective is within range of fighter-bomber aircraft, tactical air affords an excellent method of covering the withdrawal and subsequent embarkation of the supported unit. Control of airstrikes can be provided by using the naval vessel supporting the operation as a ground-to-air relay if air-naval gunfire liaison company (ANGLICO) personnel are not available.

c. Naval Superiority. Consideration must be given to naval superiority in the area of operations for this may determine the type vessel and distance from shore that the assaulting force will utilize. It may also determine the actions of the supported unit en route to the landing site, i.e., active or passive resistance, use of devious sea routes, infiltration of shipping lanes, and debarkation from minesweepers while making sweeping runs.

d. Administrative Considerations. The means of providing transportation from a rear area or marshalling area over land and sea routes are not within the capabilities of company-size units. For this reason, the type and amount of transportation available must be considered to include wheeled vehicles, oceangoing vessels, and assault boats.

(1) *Transportation.* Transportation to docks or on-shore rendezvous point with oceangoing vessels must be planned to include transportation of assault boats. Normally, 2½- or 5-ton trucks can fulfill this requirement after it is determined how many boats can be transported on each truck and how many trucks are necessary to meet troop requirements for the operation.

(2) *Vessels.* From the docks to the selected landing site, several types of oceangoing vessels are available.

(*a*) The submarine provides an excellent means of transportation for a small unit. Its characteristics enhance the chances of success due to silent, submersible movement to the landing site. However, limited space is available for personnel, equipment, and boats, and these restrictions must be considered.

(*b*) The destroyer or minesweeper provides the supported unit with greater fire support and more space for troops, equipment, and boats. The minesweeper can, in addition, launch boats while simulating a normal sweeping run if the ship is moving between 2 and 4 knots.

(*c*) The army landing craft utility (LCU) provides greater all-weather capability than the naval crafts mentioned above. If sea conditions are too severe to allow launching of boats, the LCU can beach and the troops, utilizing lifelines, can wade ashore. Embarkation can be accomplished in the same manner. This craft will accommodate 165 men, assault boats, and equipment (fig. 24).

(3) *Assault boats.* The assault boats should be considered with respect to their seaworthiness, personnel limitations, equipment limitations, and space aboard ship required to transport them to the launching site. The above considerations, when coupled with weather conditions, can render a well planned operation ineffective unless they are taken into account during planning, coordination, and rehearsal.

(4) *Signal.* Pre-coordinated signals are a necessity for a waterborne operation. They are used to request air support, artillery, and naval gunfire. If possible, signals may be used to

Figure 24. Landing craft utility.

guide the oceangoing vessel into position for launching of boats and for embarkation upon completion of the mission. Signals, even verbal, must be planned for return of friendly forces into the beachhead or into the area of the agent or partisan if this method of return is utilized. Boats should be identified within the beachhead by some visible code such as paddles in the boat in one element, in the sand, blade up in another element, and so on.

(5) *Effects of weather.* It is reasonable to expect that the hostile beach defenses will be less alert on a fogbound, rainy night, especially if the sea is rough. For this reason, consideration must be given to weather when planning the operation. The rougher the sea and poorer the visibility, the better the possibility of making a landing undetected. However, under these conditions, air support or naval gunfire cannot be relied on. In addition, if there is considerable wave action, the commander must consider the possibility of both personnel and equipment being lost in the water. The Beaufort Scale, as used by the Coast Guard for determining sea conditions, is accurate at sea. However, depending on an onshore or offshore wind, it cannot be relied on accurately in predicting surf conditions or sea conditions within ¾ of a mile from the beach.

(6) *Coordination.* Certain coordination must be effected by higher headquarters which includes partisan or special forces agent contacts, tactical air support with the Air Force, and oceangoing vessel and UDT support with the Navy, artillery support, and transportation from marshalling area to docks. After this has been accomplished, liaison officers from the other services should report to the supported unit commander

for detailed coordination with respect to air and fire support, loading plans, troop movement plans, landing plans, sea rehearsals, and the other details that must be included in order to effect a successful operation.

(7) *Techniques of operation.* There are variations of the waterborne raid; however, the planning and the execution phases require the same principles and techniques.

(*a*) *Basic tactical organization.* The basic tactical organization consists of a headquarters, an assault element, a beach security element, and a blocking element(s).

1. The assault element consists of a headquarters, an assault team, support team, prisoner team, destruction team, search team, etc. The support team is positioned, terrain permitting, prior to the actual assault to support in the event the action is prematurely triggered. In addition, it supports the withdrawal of the assault team from the objective. The headquarters, of course, provides control for the assault element.

2. The beach security element consists of a headquarters, beachmaster party, and main body. The headquarters is responsible for the overall operation of the beach security. In addition, the element leader should precede the main body of raiding unit to the beach with the beachmaster party to insure that proper agent or partisan contact is made; that the landing site is correct, and that assembly aids are correctly spotted for the landing of the main body of the raiding unit. The beachmaster party is responsible for marking the landing site with assembly aids, providing security during the landing of the beach security element, and assisting in the assembly of the main body of the raiding unit. The main body of the beach security element provides security for the blocking and assault elements during landing and embarking and for the boats during the actual raid.

3. The blocking element or elements are organized for each major avenue of approach into the objective and are responsible for an ambush type mission. This element consists of a headquarters and main body. The headquarters is responsible for control of the element. The main body should be organized into squads of fire teams, each fire team with at least one automatic weapon. It should be prepared to fight a delaying action back into the beachhead, but only after the assault mission has been accomplished.

(a) *Actions in rear or marshalling area.* The actions in this location are those normally followed in preparing for an operation. The ground tactical plan is formulated, coordination with supporting services is completed, and the raid order is issued to the element leaders. All personnel, having received the warning order, busy themselves with preparation of equipment and uniforms for their particular missions. The element leaders, after receiving the raid order, coordinate among themselves and formulate their raid orders. After issuance of the orders, a detailed inspection is conducted by the element leader. This is followed by a detailed rehearsal, on land, from ship-to-objective and return. After the element leaders' rehearsals, the raiding unit commander conducts his rehearsals on dry land. If possible, the entire operation should be rehearsed on terrain similar to the objective site. Upon completion of rehearsals, the unit is entrucked to docks or rendezvous site where it meets its supporting ocean-going vessel.

(b) *Actions at sea.* The actions of the unit at sea are similar to those of a unit in an assembly area. Equipment and weapons are cleaned and prepared, plans and rehearsals are discussed and, time permitting, personnel rest. Special attention must be given to a sea rehearsal.

(c) *Completion of rehearsal.* Upon completion of the rehearsal at sea, boats should be placed to permit cleaning of crew-served weapons, radios, and items of special equipment which will be lashed in the boats during the landing. They must, of course, come aboard in inverse order of debarking in order to avoid confusion and best support the tactical plan. Equipment, other than individual weapons and personal gear, must be lashed into boats to prevent loss when boats are placed in the water or moving through the surf.

(d) *Debarkation and movement to the beach.* The first unit to leave the ship is the beachmaster party under control of the beach security element leader. After the beach is secured, lights or other assembly aids are placed in position and the remainder of the beach security element proceeds to the beach. Individual boats are under the coxswain's control. If the sea is calm, a formation to the beach, such as wedge, is employed under the control of the element headquarters; however, in a rough sea, boats may not be able to orbit into formation and must proceed inshore individually. Under these conditions, control is difficult until elements land on their respective lights. The blocking element or elements then land followed by the assault element. By landing in this order, there is little delay on the beach. As soon as the beach security element arrives, it secures the beachhead and the boats that are following that element move in to shore. As the blocking forces land, they debark and proceed toward their assigned positions. This minimizes the time spent on the hostile shore by moving all units simultaneously. In addition, it reduces the possibility of discovery and traffic within the area continues to move freely while the assault element is moving toward the objective. The boats are located above the high watermark, preferably concealed by vegetation, with a means of identification for each element.

(e) *Movement to positions.* When all elements are in position, the assault commences. The triggering force is the assault element. However, the delay caused by the loading plan should allow the blocking elements to be in their positions prior to the assault. If this is not the case or if the assault element is prematurely discovered, the assault can still commence, because hostile reaction time will probably permit the blocking elements to move into position and execute their assigned missions. The degree of success depends upon stealth, violent execution, and rapid withdrawal of all elements. If the mission requires returning prisoners of war (POW's) or materiel to friendly lines, the blocking elements must remain in contact until the POW's or materiel are off the beach. Fire support should be utilized to the maximum to assist in breaking contact.

(f) *Reorganization and rendezvous.* All elements should employ a visual or verbal recognition signal in returning into the beach security position. Reorganization must be rapid, but complete, for the boats provide the only certain exit from the hostile shore. Re-entry should be made at a specific location and, as individuals return, the bulk of the individual's ammunition should be deposited with the beach security element. Reorganization is best effected at the boats, for they can be readily identified as to location. As each element reports to the raiding unit commander, it is directed to move to the ship. The beach security element remains in position until all other elements are in the water, calls for fire on its own position, and departs the beach. Boats are unloaded as they arrive at the ship without regard to tactical order. Reports are rendered by coxswains to the element leaders.

(g) *Alternate plan.* In the planning phase the commander must plan for the possibility of elements or individuals becoming separated or wounded and not returning to the beach. He should therefore plan for an inland assembly point to which personnel can infiltrate by teams or buddies, reorganize, and proceed overland to friendly lines. This assembly point should have all desirable characteristics of an assembly area and should be well away from the scene of action. The route of friendly lines must be included in the order and re-entry must be coordinated with friendly frontline units.

60. River Navigation

This paragraph furnishes guidance to commanders confronted with the conduct of a small unit river navigation operation. It is assumed that necessary organizational changes have been effected, because of the location and the characteristics of the area of operation. There is no attempt to outline a specific river navigation training program for the commander because of the variables involved in this type of operation.

a. Training and Navigation.

(1) *River techniques.* For best results, the commander should train his unit in all phases of river navigation prior to executing any operation involving the use of waterways. Though these type operations may appear simple, certain techniques must be mastered prior to conducting waterway operations. The training should include the use of expedient craft as well as craft designed for river navigation. This will insure the flexibility required for Ranger missions.

(2) *Basic fundamentals.*

(*a*) Basically, stretches of water are known as the curve and the reach. The curve is a turn in the river course. The reach is the straight portion of river between two curves.

(*b*) In general, the greatest velocities of current and the steepest gradients are found nearer the source of the river. Velocities may vary at all points of the river within short stretches or between points across a channel. Flow is swiftest where the channel is constricted and is slowest where the stream spreads out broad and shallow. In a meandering stream, centrifugal force throws the water to the outside of curves so that the deepest water is normally near the outside bends. For this reason, the craft coxswain should stay on the outside of curves. The outside of curves is located where the high bank is and are an indication of deeper water. Sandbars and shallow water will normally be found on the inside of curves. However, in spite of this general rule, underwater obstructions remain a problem. These obstructions can be present even in the deepest channels.

(*c*) The surface of the water directly to the front should be closely watched. Lightly rippled water, where no wind is blowing, usually indicates shallow water, sandbars, or gravel bars. A long undulating wave, however, may indicate deep water and fast current. The deep water wave is formed by a combination of swift water and fast current. A smooth surface usually indicates deep water and slightly lessened velocity. A "V" in the surface of the water generally indicates an obstruction lying parallel with the direction of current. The combination of current velocity and the size of the obstruction determines the size of the "V." The "V" is only an indication of the size of that portion of the obstruction lying very near the water surface and is not indicative of the total size of the obstruction. A rolled surface, at a particular point, usually indicates an obstruction, such as a log or tree, lying perpendicular to the direction of current.

(*d*) Whenever a tributary feeds into the main body of a river or stream, a sand delta will be found. A sand delta is a triangular body of sand found at the mouth of a river or stream. The actual location and extent of the sand delta will be dependent upon the current velocity of the main stream versus the current velocity of the subsidiary stream at the angle of joining. The composition of the river bottoms and banks is also a factor. Heavy silty rivers create greater sand deltas than do light silty streams.

(*e*) The coxswain must place himself so that he can constantly see the river course and the water surface. Occupants of the boat should remain seated so as not to obstruct the view of the operator or suddenly shift the balance of the boat. Avoid ripples, boils, and other indications of disturbed water. These disturbances can throw the craft into obstructions that will tear out the bottom of the boat. It is best to remember that in cold, fast, and very silty water, such as is found in many glacial streams, a man cannot swim for long and may go down in a matter of seconds. Avoidance of sweepers by boats is of utmost importance to the safety of the boat's occupants. Sweepers are trees that have been pulled into the river or streambed by collapsing banks or are trees rooted in the streambed. Usually, the tops of these trees bob up and down on the surface. These sweepers are extremely dangerous, because a collision with one of them may cause a boat to overturn or be torn. An overturned boat in sweepers is in double danger, because the sweepers can puncture pneumatic lifevests or cause the man in the water to be caught and held underwater by branches.

(*f*) When passing from one channel into another channel which is perpendicular to the first, the boat operator should navigate at right angles to the current into which he is traveling. He should then pass on the downstream side of the perpendicular channel. Next, the operator should proceed upstream a short distance and turn back into the flow of the new current. When using powered craft, the coxswain should never go into fast water at full throttle. The motor should be throttled down to about half speed until the operator is certain of what lies ahead. It should be remembered that a river can be read much more easily when going upstream than when going downstream.

(*g*) River channels and river obstacles are signaled by nature's signs. By learning these signs and remaining alert for them, the boat operator can navigate the rivers safely and comfortably, without damage to either his boat or motor. The boat operator should be thoroughly trained before being allowed to operate the boat alone. The reason for this is that nature's signs are general in scope and experience is required to consistently and accurately interpret them.

(3) *Navigation.*

(*a*) The river must be of sufficient width and depth to accommodate the boat or craft being used. If the unit is to move a relatively short distance or to a well defined point of debarkation, no special plan or technique precedes the operation other than a hasty route selection. The length of the move, of course, increases the chance of a navigational error. Short moves require little planning. Usually, a simple plan put to mind and committed to memory will suffice. A plan of this nature usually is as indicated here: "We'll move west until we reach the bridge about 1,500 meters from here." Finding the bridge will be a relatively easy task, especially if the waterway is not confused with many branches or tributaries.

(*b*) The use of a navigator and an observer is the best method of locating your exact position while on a river or inland waterway. The organization of the crew must be such that those duties are announced and reflected in the boat loading or boat organization plan. The navigator, who is under a poncho on the floor of the boat during hours of darkness, is equipped with a map, compass, flashlight, and pencil. The poncho precludes the unintentional reflection of light or even direct showing of the light to someone, possibly an enemy, who would be on or near the waterway. The observer, usually well forward in the boat, calls turns and bends to the navigator who, with the aid of his compass, associates this outside information with his present position on the river. Thus, the observer and the navigator work together in order to stay abreast of their location and present position. This type of navigation is such that the boat can easily move to any accessible predesignated landing site. This, of course, becomes more of a requirement when on a patrol mission utilizing boats as a means of effecting infiltration of the enemy battle position.

(*c*) As an aid to navigation, a strip map can be drawn on a piece of luminous tape showing critical check points. This can best be utilized by giving this map to someone well forward in the boat. The time-distance method is also an aid to effective navigation. Here, the time from the point of origin to the first check point can be noted and compared to the entire distance to be traveled. Thus, an estimate as to the time required can be made.

b. Techniques of River Patrolling.

(1) A river patrol may be used for reconnaissance, combat, or security. Its missions are similar to those assigned to dismounted patrols. Because of their mobility, river patrols operate at greater speeds and cover greater distances than dismounted patrols. The outstanding feature of a river patrol is its ability to carry more equipment, weapons, and ammunition than can be transported by the dismounted patrol.

(2) A river patrol can: reconnoiter the front to get information of the enemy, terrain, and the route of advance; provide security to the front, flanks, and rear; maintain contact with friendly units; clear blocked waterways; seize and hold critical terrain features; and relieve or reinforce isolated units.

(3) To form a river patrol, a minimum of four personnel per boat are selected. One of these is the coxswain. He should be an experienced operator having knowledge of river read-

ing. The remaining personnel, including the boat commander, form the minimum number of personnel necessary to handle the boat in the event that poling, lining, or rowing operations become necessary. A patrol consists of not less than two boats. This provides depth, flexibility, and safety in case one boat should come under enemy fire or be swamped or swept into obstructions (fig. 25). The number of men, weapons, and boats necessary for a patrol depends on the mission. Whenever possible, tactical unity should be preserved by selecting complete squads for a patrol. When assigning personnel to boats, tactical unity should be maintained as far as possible. A commander should be assigned to each boat.

(4) The organization of boat crews should be done in a clear and concise manner, so as not to be misunderstood by the

Figure 25. The use of two boats on patrol provides depth, flexibility and safety.

personnel involved. Simply, a crew of paddlers should be assigned and a coxswain or boat leader. The size of the boat will determine the number of additional personnel to be carried. Issue craft can be used to effect a smooth navigation of any waterway. This is especially true of the plastic or inflatible assault boat variety. Of course, the exact nature of the waterway or river will determine which boat is better for the job. Combat action may require the use of expedient boats to include civilian type craft, rafts, or floats. A handmade raft can be devised to do the job adequately.

(5) The planning for a river patrol is divided into two phases. They are the preparation phase and the actual water movement. There is little difference between a river patrol and a dismounted patrol during the preparation phase. However, there are certain areas peculiar to river patrolling that must be emphasized. These are—

(*a*) When a motor is used, it must be of sufficient horsepower to negotiate the river. This is extremely important, as an underpowered motor operating in rapid water may lead to disaster.

(*b*) Inspect the boat prior to departure insuring that it is in sound condition.

(*c*) Carry sufficient spare parts to allow for repairs should a breakdown occur.

(*d*) Insure that all the boat equipment and weapons are assembled and inspected by a qualified person. Have the motor checked for running condition, gasoline, and oil.

(*e*) Rehearse signals to be used between boats. Have the radio equipment checked to insure it is operating properly and on the correct frequency.

(*f*) Select a rendezvous point by map inspection prior to departure where members of the patrol will meet should the boat be swamped.

(6) Certain missions will require that all boats have one automatic weapon mounted on the bow plate and in firing position. Insure that the ground mounts for these weapons are in the boat so that the weapons may be used in the event the patrol goes ashore. Patrol members carry their individual weapons, sufficient hand grenades, and ammunition. As with other reconnaissance patrols, there should be more than one compass and one wristwatch and at least two pairs of fieldglasses. To increase its firepower, the patrol members may substitute automatic weapons for semiautomatic weapons.

(7) Within the patrol itself, the AN/PRC–6 radio with a planning range of 1½ kilometers, voice command, and arm-and-

hand signals may be used. Oral commands will be given when the boats are within voice range of each other. Arm-and-hand signals are effective if all men are familiar with their meaning. To communicate with higher headquarters, the radio AN/PRC–10 may be used. The commander will specify if messages can be sent in the clear. The patrol leader should use his radio sparingly. The depression of the transmission button on the radio is sometimes sufficient to relay certain information. If transmission is necessary when close to the enemy, cup the hands over the transmitter and talk in a low voice. As a control measure, the commander may assign check points along the route. Aircraft used for reconnaissance may communicate directly with the river patrol by radio or by dropping messages which give early warning of enemy activity.

(8) Execution—

(*a*) Movement on the water is accomplished close to the shoreline by taking all advantage of natural concealment. Boats are kept approximately 100 yards apart. Close bunching will offer a convenient target to the enemy and could result in a pileup in the event that one boat should have an accident. The boats in the patrol should keep their relative position in the column. The two leading boats operate as a team in moving from one observation point to another. Sharp bends in the river are frequent and obscure the river ahead. Rivers that deny observation beyond the turn are dangerous. Members of one of the two lead boats should go ashore to reconnoiter the river beyond the curve, if vision around the bend is obscured. The automatic weapons from the first boat are used to cover the advance of the investigating personnel. When the shore reconnaissance party determines that the area is clear, they signal the remaining boat to move forward. Each boat normally maintains visual contact with the boat to its front.

(*b*) Each member in the boat is assigned a specific direction in which to observe and is responsible for providing security in that direction. Smoke or footprints along the banks of the river are indications to the patrol that the enemy may be present. Water fowl suddenly alarmed and flying toward the patrol will give an indication of something ahead in the river.

(*c*) One member of each boat is appointed as an air guard. If any enemy observation aircraft is sighted, warning is passed to the other boats over the AN/PRC–6 radio and the boats are immediately maneuvered close to shore and concealed. The boat operator cuts the motor and the boat comes to a stop. Personnel sit quietly in the boats with heads lowered until the all-clear is sounded. It may become necessary to beach the boat should a hostile aircraft make a firing pass. Beaching procedure under these circumstances differs from the normal in that the boat is immediately headed toward shore regardless of direction of travel. This is especially dangerous when travelling downstream.

(*d*) If your mission requires you to fight, you should take aggressive action. For example, suppose the leading boat comes under direct enemy fire. By radio or signal, the patrol leader is notified of the situation. Personnel in the two lead boats immediately beach and place themselves in a firing position to return the enemy's fire. Using all available cover and concealment, the patrol leader moves forward and goes ashore. He decides how to best use his force to quickly knock out the enemy. By radio or signal, the patrol leader instructs the assistant patrol leader to beach all boats and bring the remainder of the patrol forward. While the patrol prepares to attack, a message is sent informing the commander of the situation. At least one automatic weapon is left at the boat assembly area. Boat operators protect their boats with automatic weapons and their individual arms. Upon capture or destruction of the enemy, the commander is notified and the mission is continued. It should be pointed out that the river patrol has a definite advantage over the dismounted patrol when confronted with this type situation in that additional weapons, such as mortars, recoilless rifles, and ammunition can be easily transported without undue hardship to the patrol personnel.

(*e*) Tributaries emptying into the river along the route should not be reconnoitered unless required by the mission. Many of these tributaries are not navigable and will cause damage to boats and unnecessary delay to the patrol. When islands are encountered, the patrol continues movement along the near side of the island avoiding open exposed areas.

(*f*) The patrol leader may order a halt to send messages, rest, eat, or observe. The area selected should provide cover and concealment and favor defense. All-around security is maintained. The position should afford good firing positions and enable each member to fire readily. Improvement of the position continues until the patrol departs.

(*g*) Secrecy of movement when travelling upstream is difficult when using a motor. The noise created by the motor at night can be heard for a distance of 5 miles. This is reduced to approximately 2 miles during daytime operations. Development of newer type models has reduced the distance travelled by motor noise to 500 meters during daylight hours. Travel downstream can be accomplished without the aid of a motor, thereby greatly increasing security.

(*h*) River patrolling at night provides a greater amount of concealment, but it is extremely dangerous. The ability of the boat operator to read the river or distinguish other navigational hazards is substantially reduced. River patrolling during periods of darkness should be avoided.

(*i*) Careful planning prior to the use of inland waterways is necessary. Many streams and rivers are not navigable and can be used only for short distances. Proper training of boat operators is essential. Adequate reconnaissance must be made of all waterways before attempts are made to use them. The course of a river or stream often provides the easiest and safest route, but may also vary from the route desired and lead to an ambush. It is important that all phases of water patrolling be examined. Careless or hasty selection will result in delays and/or loss of equipment and men.

c. Boat Selection.

(1) The performance of a small river type boat is effected by several factors: the conformation of the boat, the material from which constructed, and the weight. The type of motor, type of propeller, location of the motor and the distribution of weight in the boat effect speed and maneuverability. In general, the intended use of a boat should be the determining factor in choosing a specific type boat. For instance, the racing boat is generally flat-bottomed and built with a "step" so that the boat when racing is brought up on its "step" into three points of suspension. This cuts down drag and permits high speed. By the same token, this type boat is not as maneuverable as other types and is no good as a work boat. When considering the characteristics desired in a boat for military use, the character and velocity of the rivers on which the boat might be used has to be considered. The capacity of the boat must also be taken into consideration. Where secrecy and stealth are prime factors, rubber boats should be considered. If wooden or plastic boats are used the paddles should be wrapped with cloth to reduce noise.

(2) Nomenclautre of boats and parts are generally standard. The front is the bow and the rear is the stern. Starboard and port are the right and left sides, respectively. The bow plate is the part of the boat to which the anchor and rope are connected to the craft. The carrying handles are along the inside of the gunnel at the top of the boat and are used to lift and carry the boat.

(3) The commander in wartime has two sources from which he may obtain craft for his waterborne operation.

(*a*) Boats obtained through normal supply channels as shown in figure 26.

(*b*) Boats obtained from civilian sources in the area of operations.

Section VII. ANTIGUERRILLA OPERATIONS

61. General

Antiguerrilla operations are normally conducted by conventional units highly trained in collecting information, conducting raids, and maintaining continuous pressure on guerrilla forces. Personnel in such units are highly skilled in patrolling, night operations, evasion techniques, and land navigation. They are capable of operating in guerrilla-infested territory for extended periods.

1 Assault boat, pneumatic, 15-man

Figure 26. Boats obtained through normal supply channels.

2 Sixteen-foot plastic assault boat

3 Assault boat, plastic, M3

Figure 26.—Continued.

4 Nineteen-foot bridge erection boat

5 Boat, reconnaissance, 3-man, pneumatic

Figure 26.—Continued.

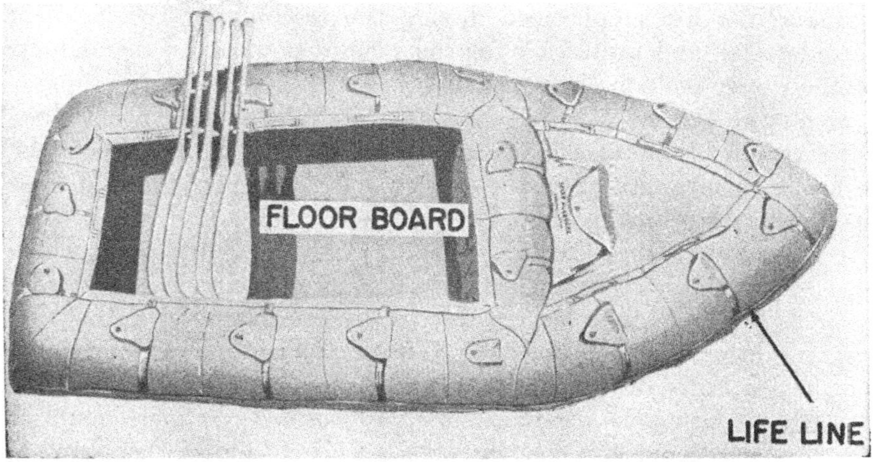

6 Five-man canvas pneumatic reconnaissance boat

Figure 26.—Continued.

62. The Guerrilla

a. Characteristics of the Guerrilla. He is very familiar with the terrain over which he operates; selects his targets; operates primarily at night, during period of reduced visibility, and during inclement weather; strives to enlist the civilian population to support his cause since he requires aid from elements not within his immediate organization for supply and conduct of operations.

b. Guerrilla Requisites for Operation.

(1) *A base* which provides security, discourages pursuit, offers adequate routes of entry and exit, provides routes to alternate bases, and is close to the area of operations.

(2) *A means of supply* for food, weapons, ammunition, and equipment. These can be gained through civilians, an external sponsor, or attacks on the enemy.

(3) *Intelligence* which enables him to plan operations or evacuate his base when endangered. Common means of gaining this intelligence include friendly civilians, monitoring enemy communications, interrogation of opposing troops, observation, surveillance, and raids.

(4) *Communications* to gain timely information and to disseminate instructions properly. Means normally employed include messengers, civilian radio equipment, and captured friendly equipment.

63. Methods of Combating the Guerrilla

To defeat the guerrilla it is necessary to destroy the security of his base, eliminate his supply, disrupt his communications, and suppress his intelligence. Personnel and units at each level of command exert an influence on the implementation of these methods. Individual responsibilities are emphasized during the training of the Soldier; consideration and protection for the general populace is mandatory. Methods used include the following:

a. Destruction of the Security of His Base. Unrelenting pressure is maintained on the guerrilla through combat and reconnaissance patrols, raids, and major unit actions. In the conduct of these operations the following should be considered:

(1) *Intelligence required to plan operations.* Some means which may furnish valuable information are reconnaissance patrols, friendly civilians, raids, air reconnaissance, enemy and friendly documents, air photography, and interrogation of captured or surrendered guerrillas.

(2) *Requirement for an immediate reaction.* Actions should be considered for contact with guerrillas along the route of march, in the objective area, and during the withdrawal. The capability to immediately react to information received of an enemy movement, action, or threat is essential. Mobility through use of vehicle, aircraft, or on foot should be evaluated. During movement of friendly units, actions to be taken upon enemy contact, to include ambush, should be planned prior to initiating movement.

(3) *Pursuit and elimination.* Complete destruction of the guerrilla force is the objective of this phase. The disbanding or retreat of a guerrilla force enables it to continue operations at a later date. Combat and reconnaissance patrols, airstrikes, and raids are examples by which continuous pressure can be applied and maintained. Special emphasis is placed on the capture or elimination of leaders.

b. Elimination of His Supply.

(1) *Eliminate sympathy for the guerrilla.* The use of troops arriving unexpectedly at villages, the gaining of information about guerrillas, their leaders and sympathizers, and the elimination of outlaw units through surprise attacks will encourage the population to resist guerrilla methods and operations. Properly disciplined troops will assist greatly in gaining the confidence of civilians during this phase.

(2) *Protection of civilians.* The appearance of friendly forces during unusual hours and at unexpected places will give confidence to civilians. An effective reaction to guerrilla attacks, the arrest of leaders, and the ambushing of local guerrilla bands will create a feeling of insecurity for the enemy while developing confidence in the population.

(3) *Security.* Establish adequate security for installations, equipment, and troops. The safeguarding of ammunition and weapons is of vital importance since the guerrilla force usually has these in limited amounts.

(4) *Supervision over movement of foodstuffs into and out of restricted areas.* These measures are established in close coordination with governing agencies. Operations by highly trained units may assist local authorities in control of these areas. Maximum coordination takes place between civilians and military personnel at all levels.

(5) *Destruction of cultivated areas.* Guerrilla forces in some localities may cultivate areas for food. No cultivated areas will be eliminated by military personnel without coordination with civil authorities and permission from the next higher headquarters.

c. Disruption of Guerrilla Lines of Communications. Methods employed include—

(1) Raids on guerrilla camps.
(2) Protection and sympathy of the populace.
(3) Security-conscious troops.
(4) Elimination of dissident elements.

d. Suppression of His Intelligence. Methods employed include—

(1) Operations of combat and reconnaissance patrols.
(2) Sympathy and protection of the populace.
(3) Security-conscious troops.
(4) Security precautions at installations.

64. Antiguerrilla Missions

a. Reconnoiter enemy territory to uncover profitable targets for airstrikes, missiles, or larger unit raids.

b. Conduct raids against guerrilla command posts.

c. Seize and hold for limited periods key ground within guerrilla-controlled territory.

d. Conduct searches and seizures to demonstrate friendly capabilities to assist civilian populace within guerrilla-infested terrain. These actions must be carried out in a manner that will not arouse resentment within the civilian population.

e. Capture or eliminate key guerrilla leaders.

f. Assist major units to maintain continuous pressure on guerrilla forces through combat and reconnaissance patrols.

65. Characteristics of Antiguerrilla Operations

a. Rigorous fire discipline.

b. Strict discipline among troops in dealing with civilians. Consideration for habits and customs of the people is enforced.

c. Raids and ambushes are continually and ruthlessly executed.

d. Movements are conducted at night when possible.

e. Familiarity with terrain is emphasized.

f. Detailed accumulation of intelligence.

g. Maximum communication capability.

h. Continuous contact is maintained with guerrilla forces through reconnaissance patrols, combat patrols, raids, and ambushes.

i. Continuous surveillance of known or suspected enemy locations through use of patrols, aircraft, observation posts, and friendly civilians.

j. Maximum use of diversions and deceptive measures.

66. Antiguerrilla Patrol Operations

a. General. The patrol is a force ideally suited to conduct antiguerrilla operations. By tailoring a force (as is done in organizing a patrol) to accomplish a specific mission, we use personnel and equipment with the greatest efficiency and insure a greater chance of successfully defeating the enemy. Highly trained units provide an effective means of eliminating the guerrilla. By employing all necessary skills, patrols can function with the flexibility and decisiveness demanded for ultimate elimination of the enemy. Furthermore, units which have attained a high skill level will be able to meet guerrilla forces on all types of terrain and in any condition of weather and destroy them in their own type of warfare. Only through keeping relentless pressure on the guerrilla can he be defeated. Contact must be made and continuously maintained. Patrol operations must be planned to facilitate hard-hitting actions which destroy the foundation of guerrilla operations—means of supply, operational bases, intelligence, and communications.

b. Characteristics of Antiguerrilla Action.

(1) Units or patrols conducting antiguerrilla warfare will use all support available in accomplishing the mission. Indirect fire may not be available because of the operating ranges of these elements. This necessitates the planning and implementation of actions which will reduce the guerrilla without this support. Tactical air support, however, will be used whenever possible.

(2) Units must be trained to operate at extended ranges for prolonged periods of time. Information contained in paragraphs 46 through 57 is applicable for these units.

(3) Targets of extreme importance are usually assigned to these highly trained units. Because of their ability to use highly developed basic skills, and the characteristics of the target, such patrols often provide the best, surest, and fastest means available to destroy guerrilla groups or command posts.

c. Characteristics of Antiguerrilla Patrols.

(1) *Organization.* Patrols dispatched on combat or reconnaissance missions will organize as prescribed for normal operations. See FM's 7–15 and 21–75. However, certain considerations in planning and conducting operations require emphasis.

(2) *Preparation of the patrol.*

(*a*) Secrecy is required throughout planning.

(*b*) Planning must be made for immediate actions to be employed when contacting the enemy to include—

1. Maximum variation in routes.

2. Detailed support such as transportation, tactical air, and aerial resupply.

3. Detailed study of terrain through aerial reconnaissance maps, and aerial photographs.

4. Use of all available means of support such as: scout dogs, indigenous personnel, trackers, and guides.

(*c*) Careful selection of personnel is made to eliminate physically unfit personnel who can jeopardize accomplishment of the mission. Personnel with slight wounds or injuries are utilized in nonpatrolling activities during periods of convalescence.

(3) *Conduct of the operation.*

(*a*) *Movement.* During movement to the objective, the following are emphasized:

1. Unencumbered movement. Do not burden a patrol with unnecessary equipment, ammunition, or weapons. Maximum mobility must exist throughout the conduct of operations.

2. Stealth. Stealth is required to gain secrecy and eliminate the enemy's possible escape prior to the time suitable offensive actions can be taken against him. The use of current intelligence often enables undetected movement through certain areas. Movement during periods of bad weather or reduced visibility will facilitate retention of stealth and thus help to attain surprise.

3. All-around security. The ability of the guerrilla force to attack from any direction requires security in every direction if base camps, installations, and rest areas are to be adequately protected. Continuous thorough policing of these areas will help secure your presence in the enemy area and thus help avoid detection.

4. Use of diversionary patrols. Additional patrols may be deliberately dispatched to areas other than the objective area for the specific purpose of deceiving the enemy as to the actual target. These patrols function in a routine manner; patrol members need not be informed the patrol is a diversionary procedure since its mission can serve additional purposes.

5. Maintenance of foot mobility. The physical and combat conditioning during training provides the basis for required foot mobility. Through use of this mobility, your patrol can move rapidly to trap the guerrilla and prevent his escape.

(*b*) *Support en route.* Units combating guerrillas should receive all available support en route to the objective. This support includes—

1. Reports from supporting elements to include civilian contacts, observation aircraft, and higher headquarters. Timely information will often determine the degree of success attainable. Without it, maximum flexibility and mobility are not achieved.

2. Indirect fire and tactical air support. When special operations are undertaken within range of indirect fire weapons or when mortars are carried with the patrol, their use is planned in detail. For extended operations, because of range and weight limitations, tactical air support may be the only support feasible. The use of either or both can create a demoralizing effect on the enemy.

3. Aerial resupply and evacuation. Distance, weather conditions, terrain, and the enemy situation will often deem aerial resupply of antiguerrilla patrols the most efficient. Morale will decrease if wounded are not evacuated promptly, so every effort is made to effect aerial evacuation. Resupply should take place during darkness and periods of inclement weather when feasible to lessen enemy detection.

4. Actions at the objective—

(a) *Reconnaissance patrols.* The use of highly trained units in accomplishing sensitive reconnaissance missions is mandatory because of the need to function undetected. A continuous flow of information must be realized if the proper offensive actions are to occur. Patrols may be required to perform area or point reconnaissance (1, fig. 27). For definitions of terms, see FM 7–15. Surveillance of critical targets can be conducted by reconnaissance patrols which halt during daylight hours to merely observe activities (2, fig. 27). Information of immediate tactical significance is reported without delay. Additional information will be reported upon return to the base camp.

(b) *Combat patrols.* Highly trained small units because of their flexibility and teamwork, provide the commander with an effective means for conducting raids and ambushes. However, small forces may be ineffective against well organized, fortified, and strongly

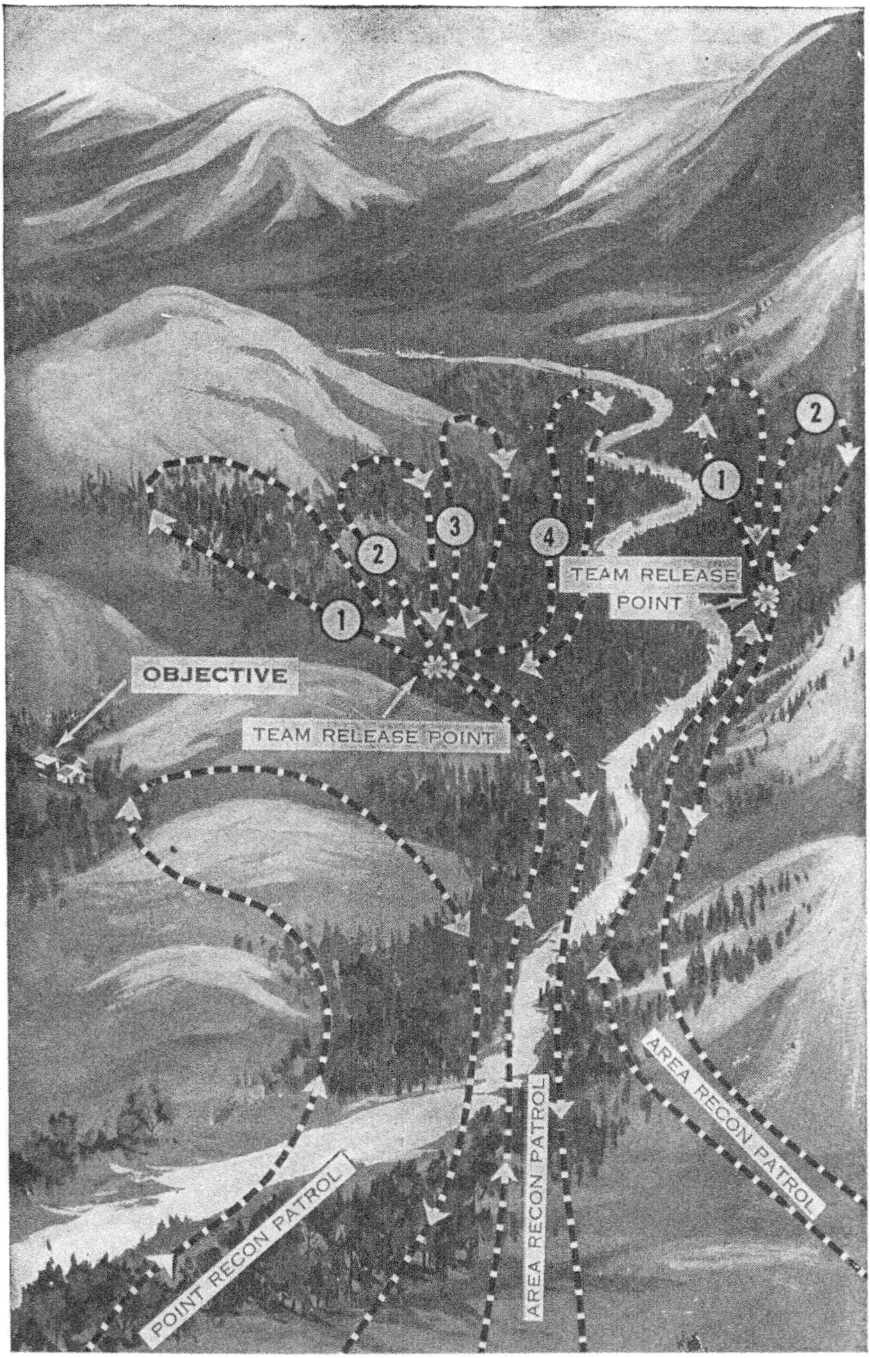

1 Area and point reconnaissance

Figure 27. Area or point reconnaissance and stationary observation.

2 Stationary observation

Figure 27—Continued.

defended positions unless supported by larger units. Large units are normally needed to conduct major offensive actions against guerrilla forces. Small units in coordination with these units can effectively assist in the destruction of a guerrilla unit. The force selected must be of sufficient size and strength to insure that it is capable of accomplishing the mission.

(*1*) *Encirclement.* Encirclement is the most effective way of fixing guerrillas in order to bring about their their destruction. Although tactical developments may dictate the use of a highly trained unit, company size or less, larger units are normally employed in encircling movements. Since a small unit may be a part of the larger encircling operation or even conduct one of its own against a small target, an explanation of encirclements is included. The patrol, if encircling the enemy, will form its elements into a perimeter when at the objective. Planning, preparation, and execution of such an operation are designed to give sudden complete encirclement which will totally surprise the guerrillas. One of the major reasons for habitual use of larger forces in encirclement operations is the requirement for support and reserve elements. It must be remembered that guerrillas may react violently in order to break through the encirclement. Probing for gaps and attacking weak points is to be expected. Movement to the line of encirclement is normally accomplished at night or during periods of reduced visibility in order to permit secrecy and surprise. Units may well be called upon to isolate and hold pockets of resistance which jeopardize the major encircling effort. Through use of encirclement, four final means of destruction can be achieved:

a. By a simultaneous tightening of the encirclement (1, fig. 28).

b. By driving a wedge through the trapped force, followed by annihilation of the guerrillas in each subarea (2, fig. 28).

c. By use of the "hammer and anvil"—establishing a holding force and driving the guerrilla against it (3, fig. 28).

d. By having a strong assault force hit the encircled guerrillas (4, fig. 28). Strongly fortified positions should be reduced by indirect fire or tactical air if exposed. However, a highly trained unit can

1 Simultaneous tightening of full encirclement

Figure 28. Final means of destruction.

2 Encirclement with wedge

Figure 28—Continued.

3 Encirclement into hammer and anvil

Figure 28—Continued.

4 **Encirclement with strong assault force**

Figure 28—Continued.

effectively ferret out and reduce the enemy within such fortifications when they are hidden. In all of these maneuvers, thorough combing of every area, including those most inaccessible, is mandatory. All personnel captured are held until proper identification is established.

(2) *Attack.* The ability to completely surprise guerrilla forces does not always exist; however, surprise should always be striven for in offensive operations. Whenever possible, attacks are launched against the guerrilla to maintain continuous pressure. Maintenance of contact and timely intelligence is essential to the mounting of any attack. A small unit may conduct such operations by itself or may team up with a larger unit. In either case, the mobility, teamwork, and skill level of the small force must be capitalized upon. Attacks can be carried out in the following manner:

a. By forcing the enemy to withdraw into established killing zones (5, fig. 28).

b. By use of the double envelopment (6, fig. 28).

(3) *Pursuit.* Here the small highly trained unit is well suited to the operations demanded. Small forces with experienced trackers using speed, stealth, decisiveness, and coordination are tailored to fit the tactical needs. Ambushes and attacks upon fleeing guerrillas can be efficiently accomplished. The use of aerial observation, aerial transport, and tactical air support greatly enhance the ability of these forces to assist in the total destruction of a guerrilla force (7, fig. 28).

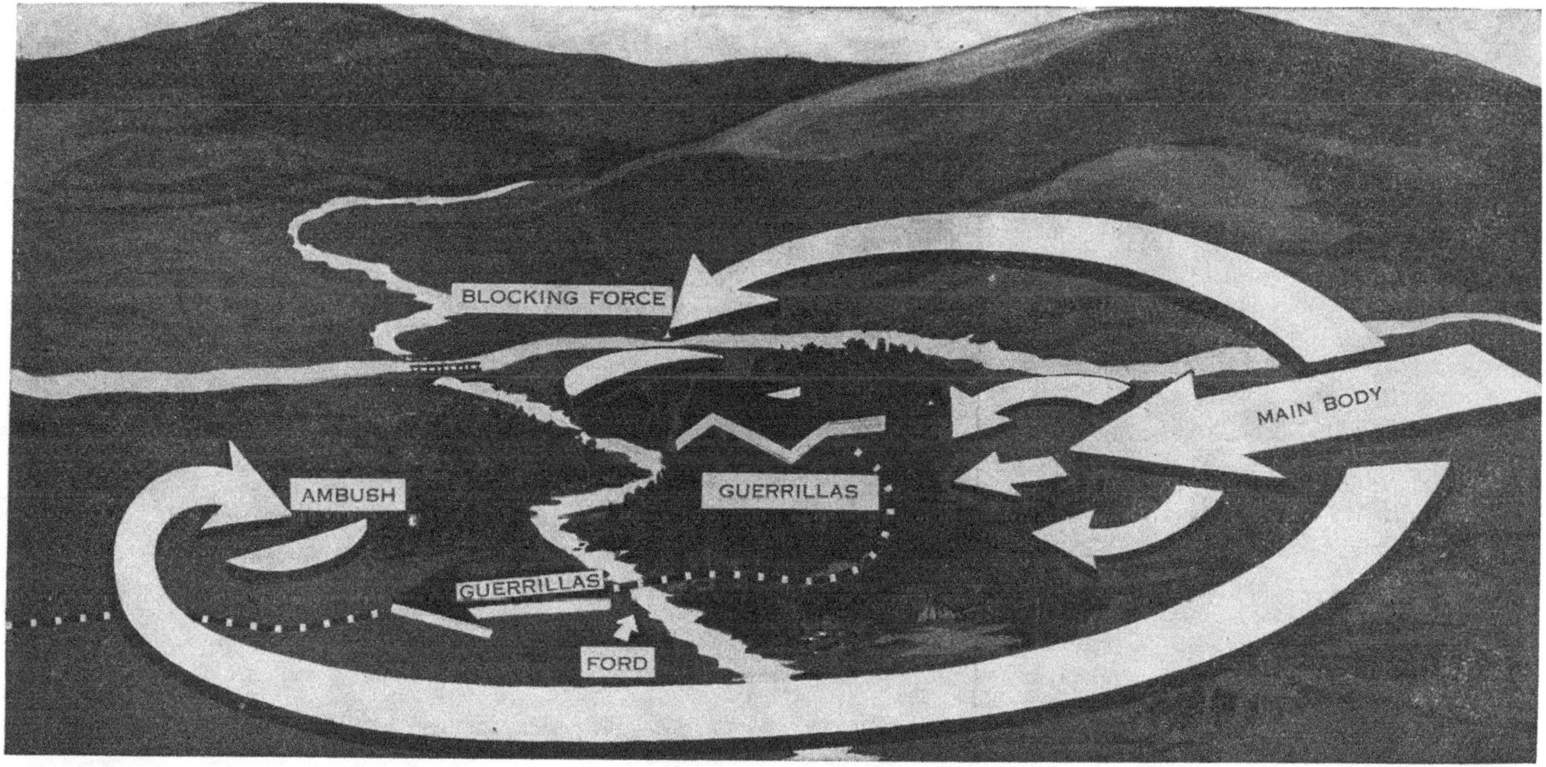

5 **Attack driving into killing zones**

Figure 28—Continued.

6 Attack by double envelopment

Figure 28—Continued.

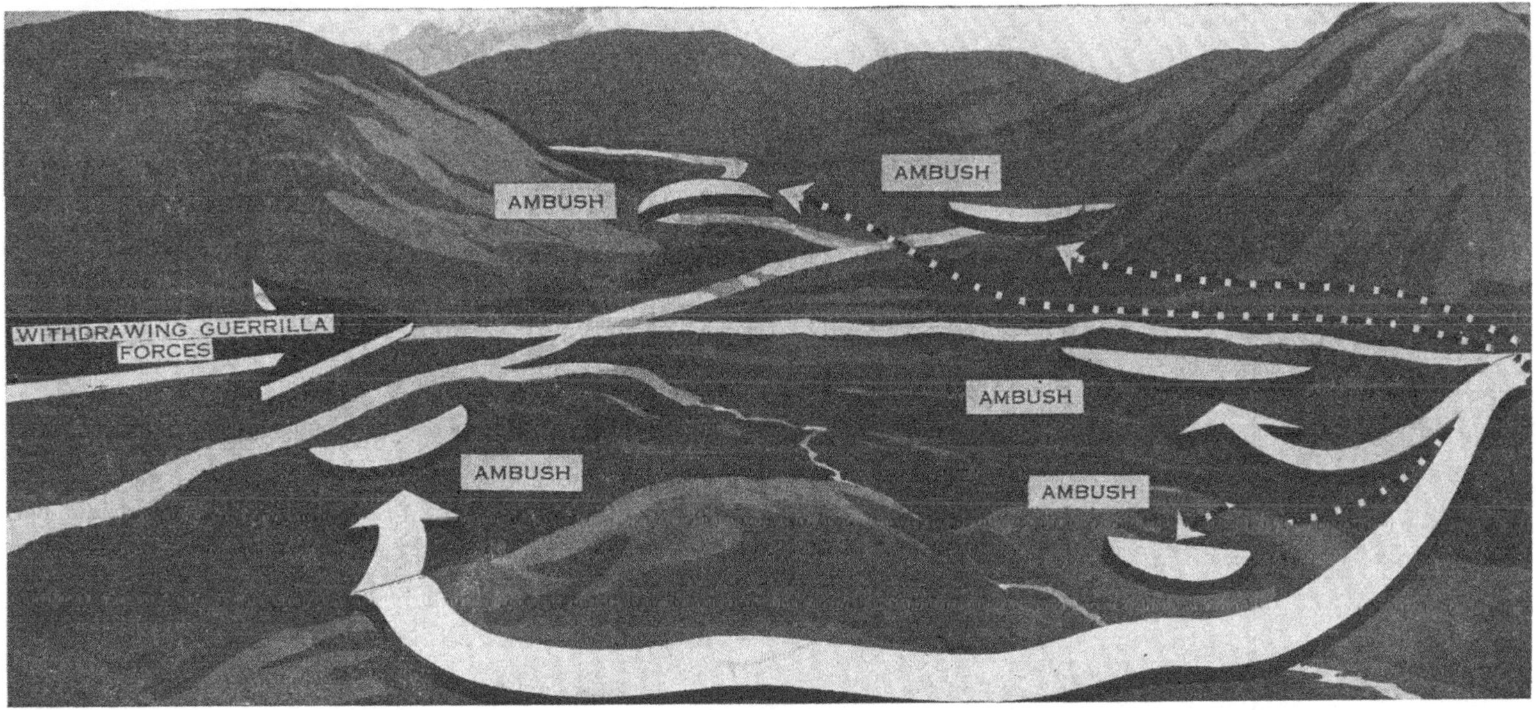

7 Small units and patrols dispatched to counter pursued forces

Figure 28—Continued.

APPENDIX I

REFERENCES

AR 350–30	Code of Conduct
AR 350–225	Survival, Evasion, and Escape Training
DA Pam 21–71	The U.S. Fighting Man's Code
DA Pam 21–81	Individual Training in Collecting and Reporting Military Information.
DA Pam 30–101	Communist Interrogation, Indoctrination, and Exploitation of Prisoners of War.
DA Pam 108–1	Index of Army Motion Pictures, Filmstrips, Slides, and Phono-Recordings.
DA Pam 355–6	Officer's Call: Command and the Code Never Surrender
DA Pam 355–51	I am an American Fighting Man
FM 5–15	Field Fortifications
FM 5–20	Camouflage, Basic Principles and Field Camouflage
FM 5–25	Explosives and Demolitions
FM 5–31	Use and Installation of Boobytraps
FM 7–10	Rifle Company, Infantry and Airborne Division Battle Groups
FM 7–15	Infantry, Airborne Infantry and Mechanized Infantry, Rifle Platoons and Squads.
FM 20–32	Land Mine Warfare
FM 20–33	Ground Flame Warfare
FM 21–6	Techniques of Military Instruction
FM 21–20	Physical Training
FM 21–26	Map Reading
FM 21–31	Topographic Symbols
FM 21–75	Combat Training of the Individual Soldier and Patrolling
FM 21–76	Survival
FM 21–77	Evasion and Escape
FM 21–77A	Evasion and Escape (U)
FM 21–150	Hand-to-Hand Combat
FM 22–100	Military Leadership
FM 23–25	Bayonet
FM 23–30	Grenades and Pyrotechnics
FM 30–5	Combat Intelligence
FM 30–7	Combat Intelligence Battle Group, Combat Command, and Smaller Units.
FM 30–101	Aggressor, the Maneuver Enemy
FM 31–15	Operations Against Irregular Forces
FM 31–20	Special Forces Operational Techniques (U)
FM 31–21	Guerrilla Warfare and Special Forces Operations
FM 31–25	Desert Operations
FM 31–30	Jungle Operations
FM 31–70	Basic Cold Weather Manual
FM 31–71	Northern Operations
FM 31–72	Mountain Operations

FM 57-35	Airmobile Operations
FM 57-38	Pathfinder Guidance for Army Aircraft
FM 110-115	Amphibious Reconnaissance; Joint Landing Force Manual
TF 7-1263	The Bayonet Fighter
TF 7-1475	Military Rock Climbing, Technique of Climbing
TF 7-1480	Military Rock Climbing, Movement of Combat Units
TF 7-1518	Summer Mountain Movement and Bivouacs
TF 7-1714	Reconnaissance Patrol
TF 7-1750	Combat Patrols
TF 21-2720	Code of the Fighting Man
TM 5-279	Suspension Bridges for Mountain Warfare
TM 9-1900	Ammunition, General
TM 9-1946	Demolition Materials
TM 21-200	Physical Conditioning
TM 57-210	Air Movement of Troops and Equipment

APPENDIX II

TRAINING SCHEDULES

RECAPITULATION OF SUBJECT HOURS

	1 week	*2 weeks*	*3 weeks*	*Ref* App.
	Number of hours			
Aerial resupply and air landed operations	0	3	3	III
Ambush and road block techniques	0	1	1	III
Bayonet	0	1	1	III
Clandestine assembly areas	0	1	1	III
Cliff assault techniques	0	0	1	III
Combat formations	0	1	1	
Commander's time	3	5	5	III
Confidence tests	2	2	4	III
Current enemy situations	0	1	2	III
Demolitions	0	3	7	III
Escape and evasion	0	1	2	IV
Field training exercises	27	73	111	IV
Critiques	2	6	10	III
Hand to hand combat	7	7	7	III
Inspections	2	3	5½	III
Intelligence	0	1	1	III
Map and aerial photo reading	0	6	6	
Use of map and compass	1	1	1	
Night compass course	9	9	9	III
Mountain techniques and expedients	6	8	9	III
Orientation	1	1	1	III
Patrol plans, orders, and techniques	6	7	7	III
Physical conditioning	4	8	10	III
River crossing expedient	0	0	2	III
Survival	0	1	3	III
Total hours	70	150	210½	

Day / Hour	1st	2d	3d	4th	5th	6th	7th	8th	9th	Night	Remarks
1st	Orientation (Ranger)	Hand to hand combat		Patrol planning orders and techniques		Use of map and compass	Hand to hand combat	PT	Night compass course to 0100 hours		(1) (2) (5)
2d	Hand to hand combat	Introduction to mountain expedients and rope work		Practical work, mountain techniques and expedients				PT	Hand to hand combat	Mountain training films	(1) (2) (3)
3d	Hand to hand combat	Patrol planning, orders, and techniques (walk through) Problem #9				Day and night reconnaissance patrol to 2400 hours Problem #1					(2) (3) (4)
4th	Critique	PT	Hand to hand combat	PT mass games	Night combat patrol to 0600 hours (raid) Problem #2						(2) (3) (4)
5th	Critique	Confidence tests		Close bivouac area			Commander's time				(3)

(1) Physical training may be integrated into movement between class sites.
(2) C-rations meal.
(3) Inspection conducted 30 minutes prior to first hour of instruction.
(4) Warm-up exercises conducted prior to instructional periods.
(5) Unit moves to field 0500 hrs and establishes bivouac area.

Figure 29. One week Ranger training schedule.

<table>
<tr><th>Day / Hour</th><th>1st</th><th>2d</th><th>3d</th><th>4th</th><th>5th</th><th>6th</th><th>7th</th><th>8th</th><th>9th</th><th>Night</th><th>Remarks</th></tr>
<tr><td>1st</td><td>Orientation (Ranger)</td><td colspan="2">Hand to hand combat</td><td colspan="5">Map and aerial photograph reading</td><td>PT</td><td></td><td>(5)
(1)</td></tr>
<tr><td>2d</td><td>Map reading (exam)</td><td>PT</td><td>Hand to hand combat</td><td colspan="3">Patrol planning, orders and techniques</td><td>Use of map and compass</td><td colspan="3">Night compass course to 2400 hours</td><td>(2)
(3)</td></tr>
<tr><td>3d</td><td>Hand to hand combat</td><td colspan="2">Introduction to mountain expedients and rope work</td><td colspan="5">Practical work, mountain techniques and expedients</td><td>PT</td><td>Training film (mtn tech)</td><td>(3)
(2)</td></tr>
<tr><td>4th</td><td colspan="2">Hand to hand combat</td><td colspan="3">Aerial resupply and air landed operations</td><td colspan="3">Demolitions Problem #8</td><td>PT</td><td></td><td>(3)</td></tr>
<tr><td>5th</td><td>Hand to hand combat</td><td colspan="2">Mass games</td><td>Combat formations</td><td colspan="4">Patrol planning, orders, and techniques
Problem #9 (walk through)</td><td>PT</td><td>Training film (escape and evasion)</td><td>(3)</td></tr>
<tr><td>6th</td><td>PT</td><td>Intelligence</td><td>Survival</td><td colspan="7">Day and night reconnaissance patrol to 2200 hours Problem #1</td><td>(3)</td></tr>
</table>

7th	Critique		Open		Commander's time	Current enemy situation	PT	
8th	PT	Bayonet	Ambush and road-block tech-niques	Ambush patrol to 0600 Problem #3			(3)
9th	Critique		Clandestine assembly area	Survival	Long range patrol		(2) (4)
10th	Hold key enemy installation to 0600 hours Problem #4						(2)
11th	Critique		Confidence tests	Close bivouac area	Commander's time		

(1) Physical training may be integrated into movement between class sites.
(2) C-rations meal.
(3) Inspection conducted 30 minutes prior to first hour of instruction.
(4) Warm-up exercises conducted prior to instructional periods.
(5) Unit moves to field 0500 hours and establishes bivouac area.

Figure 30. Two week Ranger training schedule.

Day / Hour	1st	2d	3d	4th	5th	6th	7th	8th	9th	Night	Remarks
1st	Orientation (Ranger)	Hand to hand combat		Map and aerial photograph reading					PT		(1) (5)
2d	Map reading (exam)	PT	Hand to hand combat	Demolitions Problem #8			Use of map and compass	Night compass course to 2400 hours			(2) (3)
3d	Hand to hand combat	Patrol planning orders and techniques			Demolitions Problem #8				PT		(3)
4th	Introduction to mountain expedients and rope work		Practical work, mountain techniques and expedients						PT	Training film (mtn tech)	(3)
5th	Hand to hand combat		Aerial resupply and air landed operations			Survival			PT		(3)

6th	Hand to hand combat	Intelligence	Escape and evasion	Combat formations	Patrol planning, orders, and techniques Problem #9 (walk through)		PT	Training film (escape and evasion)	(3)
7th	PT	Bayonet	Open			Commander's time	Current enemy situation		(3)
8th	Mass games		Day and night reconnaissance patrol to 2200 hours Problem #1						(3)

See notes on page 161.

Figure 31. Three week Ranger training schedule.

Day / Hour	1st	2d	3d	4th	5th	6th	7th	8th	9th	Night	Remarks
9th	Critique		Night raid on enemy lines to obtain prisoners to 0300 hours Problem #5								(4)
10th	Open	Critique		Open					Ambush and roadblock techniques		(4) (3)
11th	Ambush patrol to 0600 hours Problem #3										
12th	Open	Critique		Confidence test		Open					(4) (3)
13th	Cliff assault tech-niques	River crossing ex-pedients		Raid against enemy guerrilla camp to 0400 hours Problem #6							(4) (3)
14th	Critique		Open						Current enemy situation		(4)

15th	PT	Clandestine assembly area	Long range raid to seize and hold key enemy Problem #4				(3)
16th	Installation						
17th	Critique		Confidence tests	Close bivouac area	Commander's time		(1)
18th							

(1) Physical training may be integrated into movement between class sites.
(2) C-rations meal.
(3) Inspection conducted 30 minutes prior to first hour of instr cti
(4) Warm-up exercises conducted prior to instructional periods.
(5) Unit moves to field 0500 hrs and establishes bivouac area.

Figure 31—Continued.

623989°—62——11

APPENDIX III

LESSON SUBJECT OUTLINES

1. Aerial Resupply and Airmobile Operations

a. Objective. To familiarize Soldiers with the procedure for aerial resupply and airmobile operations. This procedure includes selection and marking of a drop zone or loading zone; and preparing patrol order annex covering air resupply, recovery and distribution, air movement, and air-ground communication procedure.

b. References. FM's 31–20, 57–35, 57–38 and TM 57–220.

c. Requirements.

(1) *Cadre.* One principal instructor.

(2) *Training aids.* Blackboard, chalk, eraser.

d. Instructor's Notes.

(1) Aerial resupply, evacuation, or movement planned and coordinated in advance.

(2) Air safety for troop movement.

(3) Aircraft loading plans and techniques should be studied and rehearsed.

(4) For patrol evacuation by helicopter, maintain all-round security for the landing zone until the last man is aboard.

(5) The loading manifest includes name, plane number, and seat in craft for each person.

(6) Technique of recovering supplies delivered by air; the selection, preparation, and operation of landing sites, drop zones, and loading sites must be understood by team or unit leaders.

(7) The information that must be transmitted to the aircraft flight leader is as follows:

(*a*) *Fixed-wing landing zone.*

1. Landing instructions.

(a) Vector (magnetic azimuth from checkpoint to LZ).

(b) Enemy situation (as affects aircraft).

(c) Wind (direction).

(d) Land (magnetic azimuth for aircraft to land).

(e) Traffic pattern (given only when traffic is not left-hand).

(f) Condition of runway (sod, gravel, etc.).

(g) Field elevation.

(h) Call base (when pilot reaches base leg of field aircraft pattern).

2. When aircraft calls base.

(a) Clear to land.

(b) Parking instructions.

3. Takeoff instructions.

(a) Taxi instructions.

(b) Clear to take off.

(*b*) *Helicopter landing zone.*

1. Vector.

2. Enemy situation.

3. Wind.

4. Landing site azimuth (if requested or necessary).

5. Land.

6. Clear to land.

(8) Use correct message format for fixed-wing resupply:

(*a*) Vector.

(*b*) Enemy situation.

(*c*) Drop formation and drop altitude.

(*d*) Field elevation.

(*e*) Maintain ---------------------- until I have you in sight.
(course and/or elevation)

(*f*) Descend to --------------------------.
(elevation)

(*g*) Steer right (as pilot is facing).

(*h*) Steer left (as pilot is facing).

(*i*) On course.

(*j*) Standby.

(*k*) Execute, execute, execute.

e. Aerial Resupply Annex to Patrol Order.

(1) *Enemy situation.*

(2) *General plan.*

(3) *Specific duties and coordinating instructions.*

(*a*) Security team.

(*b*) Signal team.

(*c*) Recovery team.

(*d*) Distribution team.

(*e*) Rear security team.

(*f*) Routes to and from drop zone (1, fig. 32).

(*g*) Time of drop.

(*h*) Action upon enemy contact.

(*i*) Alternate plans.

(4) *Administration and logistics.*

(*a*) Contents of drop.

(*b*) Method of distribution.

(5) *Command and signal.*

(*a*) Location of patrol leader.

(*b*) Identification of drop zone.

(*c*) Communications.

(*d*) Alternate signals.

f. Aerial Resupply Along Patrol's Route (1 and 2, fig. 32).

g. Aerial Resupply Within Defensive Perimeter (3, fig. 32).

2. Ambush and Roadblock Techniques

a. Objective. To teach the Soldier the principles involved in the selection, planning, organization, and establishment of a roadblock, to include an ambush and the application of these principles in various situations.

b. References. FM's 5–15 and 7–15.

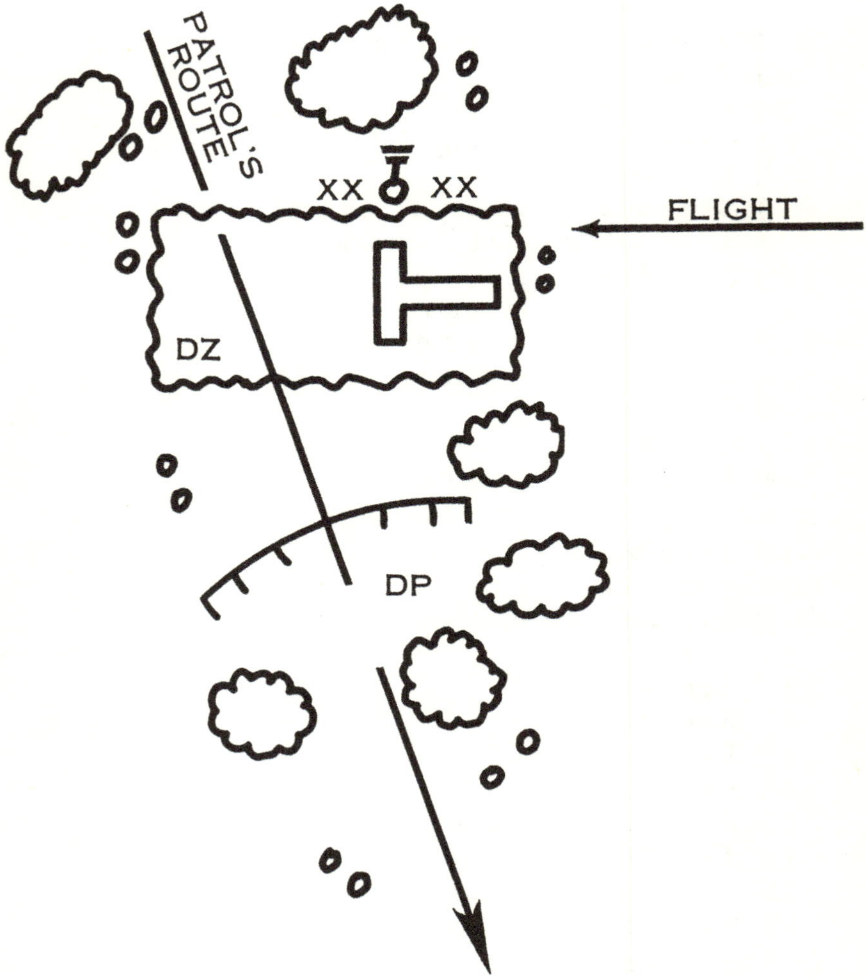

1 Organization before airdrop

Figure 32. Aerial resupply.

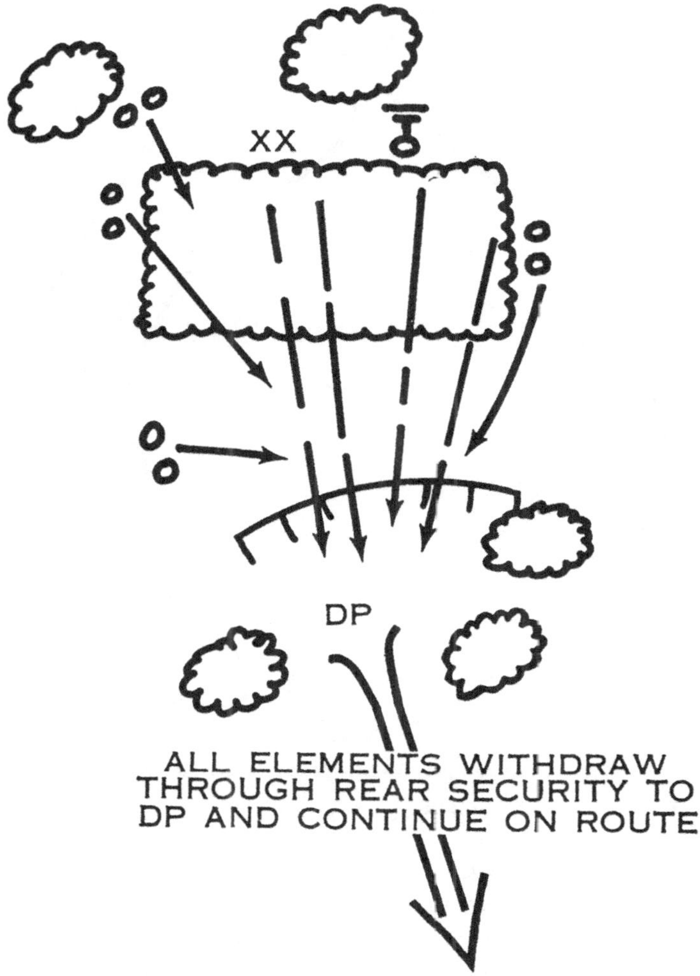

2 Action after airdrop

Figure 32—Continued.

c. Requirements.

(1) *Cadre.* One principal instructor.

(2) *Training aids.* Blackboard, chalk, eraser, sandtable.

d. Instructor's Notes.

(1) *Ambush.* An ambush is a surprise attack upon a moving or temporarily halted enemy with the mission of killing or capturing the enemy and destroying his equipment. The ideal ambush traps the enemy and allows none to escape.

(2) *Organization and tasks of an ambush patrol.*

(*a*) *Security element.* This element is responsible for early warning and protecting the flanks and rear of the assault element while it is moving into position, executing the ambush, or during its withdrawal. Close coordination between the security and assault elements is essential.

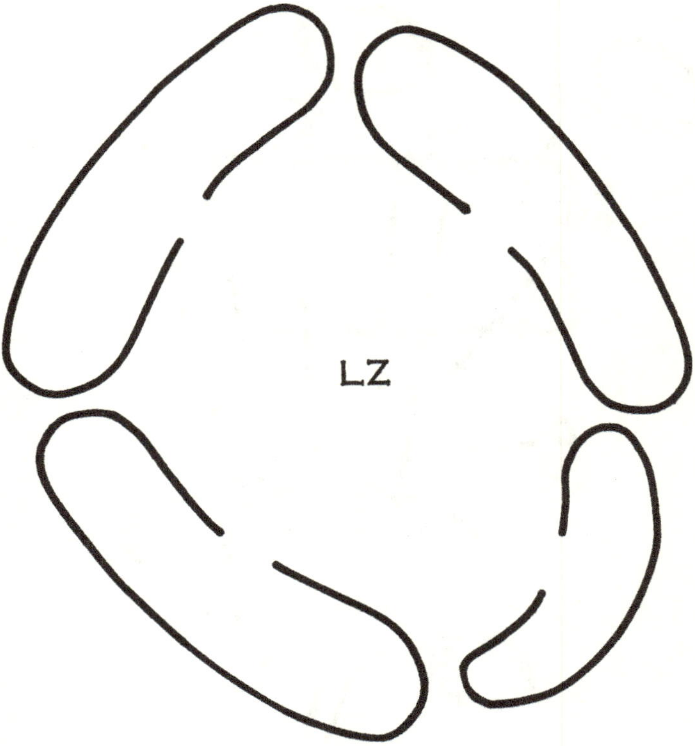

3 Aerial resupply within defensive perimeter

Figure 32—Continued.

(*b*) *Assault element.*

1. Destroys or captures the enemy at the ambush site.

2. Conducts a search to insure that all of the enemy have been killed or captured, and that the equipment has been destroyed.

3. May function under the direct supervision of the patrol leader.

4. Coordinates with the security element and establishes signals for the withdrawal.

(*c*) *Support element.*

1. May function as a team and be organic to the assault element.

2. Normally provides the base of fire at the ambush site.

(3) *Factors necessary for a successful ambush.*

(*a*) Patience on the part of each member.

(*b*) Adequate intelligence information.

1. Size of the enemy unit.

2. Type of security he employs.

3. Type weapons he possesses and how he employs them.

4. Time he uses roads and trails.

(*c*) Camouflage discipline.

(*d*) Each individual *must* be thoroughly familiar with the plan.

(4) *Characteristics of an ideal ambush site.*

(*a*) Concealment until time of attack.

(*b*) Good observation and fields of fire.

(*c*) Route affording a rapid withdrawal.

(*d*) An area that restricts the enemy's movement to one flank.

1. A natural restricted flank, i.e., lake or a cliff.

2. Use of mines, Claymore weapons, barbed wire, or automatic weapons fire to establish a restricted flank where necessary.

(*e*) An assembly area on the route of withdrawal to reassemble the patrol elements.

(5) *Execution of ambush plan.*

(*a*) The security element moves into position first and is followed by the assault element which, if necessary, sets up an artificially restricted flank.

(*b*) The assault element arranges to seal off the front and rear of the ambush site with fire.

(*c*) Physical obstacles are not placed across the enemy approach because this would eliminate surprise. However, concealed demolitions placed under a bridge, in a culvert, or on nearby trees may be triggered with the assault fires.

(*d*) The assault element opens fire on a prearranged signal established and controlled by the patrol leader.

(*e*) The assault element delivers a heavy volume of accurate fire until the cease fire signal is given.

(*f*) A search team from the assault element insures the mission is completed.

(*g*) The assault element followed by the security element bebegins a rapid withdrawal to the rallying point.

(6) *Characteristics of a good rallying point.*

(*a*) Near ambush location and in general direction of friendly lines or clandestine assembly area.

(*b*) Easily recognizable by day or night.

(*c*) Affords cover from small arms fire delivered from the vicinity of the ambush site.

(*d*) Can be defended for a short period of time.

(7) *Actions at the rallying point.* Reorganization is completed as rapidly as possible and the withdrawal phase is executed.

(8) *Defense against an ambush.*

(*a*) Employ proper security at all times when moving in a combat area.

1. Avoid using trails and roads that are utilized by the enemy.

2. Change routes and times of movement frequently.

3. Reconnoiter likely ambush sites prior to moving an entire unit into an area favorable for an ambush.

(*b*) If ambushed, every individual should open fire in the direction of the enemy and, on command, assault in that direction. Ambushed force must take full advantage of any enemy weak points during the breakout. The enemy selects ambush site similar to ours. Restricted flanks and planned fires cover the entrance and exit to a site. Movement toward the enemy is the most likely avenue of escape. If the mission is to destroy all enemy personnel, return and destroy the ambush element; if not, effect reorganization and continue on the assigned mission.

(9) *Roadblocks.*

(*a*) Roadblocks are an essential element of the barrier plan. The roadblock's mission is to prevent or hinder enemy movement past a point or area along a road. It usually incorporates obstacles covered by fire. The size of the force defending the roadblock may vary from a few men to a reinforced rifle company.

(*b*) In defensive operations, roadblocks are employed to the front, flanks, and rear of friendly units. In offensive operations, roadblocks may be used to protect the flanks of advancing columns and in the enemy's rear to prevent his withdrawal or reinforcement.

(10) *Barriers.*

(*a*) A barrier is a coordinated series of natural and artificial (manmade) obstacles placed across expected avenues of approach and linked together in linear form.

(*b*) Barriers are planned for and executed by all echelons of command unless restricted by specific order of a higher headquarters or by lack of authority to employ certain types of minefields. The development of extensive barrier systems of major tactical significance will usually be directed by corps or higher headquarters based on the overall concept of the major command.

(11) *Obstacles.*

(*a*) An obstacle is any natural terrain feature, condition of soil or climate, manmade object or works which hinders, slows or stops an advance. Examples of obstacles are rivers, swamps, ditches, abatis, craters, and minefields.

(*b*) Obstacles should be camouflaged or employed in such a way that they present surprise to the enemy. Obstacles are covered by fire and observation to make them difficult to breach or overcome.

(12) *Desirable characteristics of a roadblock site.*

(*a*) *Blocks the avenue of approach.* The roadblock is positioned so that it is difficult to bypass; however, in certain cases a higher commander may desire to violate this in order to canalize the enemy.

(*b*) *Takes advantage of natural obstacles.* Roadblocks along the side of a steep hill, across streams, marshes, ravines, etc., are examples of taking advantage of natural obstacles. Artificial obstacles can then be easily constructed to reinforce, supplement, and tie in with the already present natural obstacles. This creates an effective obstacle with minimum effort.

(*c*) *Easily defensible.* The troop position selected should provide for all-round defense and cover the obstacle and its approaches by fire observation to prevent its being breached or overcome.

(*d*) *Gains surprise.* It is desirable that the enemy be surprised in order to inflict maximum casualties and confusion upon him. Troops and artificial obstacles should be concealed from the enemy until it is too late for him to react effectively. Mines and demolitions and cratering charges are examples of effective, easily concealed obstacles. These also have the advantage of being quickly armed and unarmed as the situations may suddenly dictate in fluid operations. The location of obstacles around a sharp bend in the road, just over the crest of a hill, or where the road passes through a heavily wooded area may also be used to help gain surprise.

(*e*) *Good routes to the rear.* This not only facilitates resupply, but permits a rapid withdrawal to the rear. A slow withdrawal or bottleneck during the withdrawal of the roadblock force can lead to its destruction by a pursuing enemy.

(13) *Planning.* In planning the construction and defense of a roadblock, the normal troop-leading steps are accomplished. After considering his mission, the roadblock commander—

(*a*) Makes a terrain analysis from map and ground reconnaissance.

(*b*) Estimates material and special equipment needed to construct the obstacles and troop position.

(*c*) Makes tentative tactical plans to include—

1. A detailed fire plan and fire control measures which include specific sectors of fire and a signal for opening fire.

2. Troop and weapons locations that can place effective fire on the flanks, on the obstacle and its approaches, and can prevent the enemy from deploying around the obstacle, but not within hand grenade range of the obstacle and its approaches.

3. A withdrawal plan that includes covered and concealed routes of withdrawal, indirect fires to break contact with the enemy and to cover the withdrawal, and plans for the occupation of successive positions to cover the withdrawal. Rally points are selected on successive positions to the rear to regain control.

4. Completes his plan, issues orders, and proceeds to organize, prepare, and occupy the position.

(14) *Preparation.* In preparing a roadblock position, the roadblock commander establishes a priority of work. A recommended priority is—

(*a*) *Establish security.* All-round security is established not only to provide early warning, but also to provide coordination for passing friendly elements.

(*b*) *Simultaneous construction of the obstacle, clearing fields of fire, and establishment of communications to the rear.* The placement of blocking obstacles, such as mines, is not completed until a specific time or condition has been met. This allows the later use of the road and permits the rapid withdrawal of friendly forward forces.

(*c*) *Construction of emplacements, shelters, and secondary obstacles and construction or improvement of existing routes of communication.* These may take place simultaneously and are phased in as (*b*) above has been accomplished.

(15) *Conduct of a roadblock defense.*

(*a*) As the enemy aproaches the roadblock, the security warns the roadblock commander. On order, the security is withdrawn, without becoming engaged, along predesignated routes to the defensive area.

(*b*) Since surprise is a desirable aspect of roadblock defense, the defending force does not want to prematurely disclose its position and waits for the enemy to reach the blocking obstacle before reacting. At that time maximum fire is brought on the obstacle and its approaches, to include surprise artillery fire along the road behind concealing terrain features.

(*c*) As the enemy deploys to assault the position, direct and indirect fires are shifted to likely enemy assembly areas, avenues of approach into the position, and the flanks.

(*d*) Before the enemy is in a position to overwhelm the defending force, permission should be requested to withdraw. Secondary obstacles, artillery and mortar fires, and the occupation of successive positions to the rear are used to cover the unit's rapid withdrawal. Rally points are selected on successive positions to the rear to regain control.

3. Bayonet

a. Objective. To instill in the Soldier the spirit of the bayonet and to maintain the aggressiveness of the unit being trained. The scope of instruction will depend on previous bayonet instruction received by the Soldier.

b. References. FM 23–25 and TF 7–1263.

c. Requirements.

(1) *Cadre.*

(*a*) One principal instructor.

(*b*) One assistant instructor demonstrator per thirty Soldiers.

(*c*) Support.

1. One sound equipment operator.

2. One aidman.

(2) *Vehicle.* One ambulance.

(3) *Training aids.*

(*a*) PT platform.

(*b*) One sound equipment set with four speakers.

d. Instructor's Notes (one hour review).

(1) Explain, demonstrate, and conduct practical work in:

(*a*) *Basic positions.* Guard, short guard.

(*b*) *Movements.* High port, whirl, thrusts, and withdrawals.

(*c*) *Series.* Long and short thrust series; horizontal and vertical butt stroke series.

(2) Emphasize proper body position, footwork, balance, and spirit of the bayonet.

(3) Have assistant instructor correct errors.

4. Clandestine Assembly Areas

a. Objective. To familiarize the Soldiers with the selection and occupation of a clandestine assembly area.

b. Requirements.

(1) *Cadre.* One principal instructor.

(2) *Training aids.* Blackboard, chalk, eraser.

c. Instructor's Notes (clandestine assembly area).

(1) *Map study.* Select an area with the following characteristics:

(*a*) Cover and concealment.

(*b*) Access routes.

(*c*) Away from routes that offer natural lines of drift for aggressor forces.

(*d*) Sufficiently close to area of primary operation to afford maximum use of time and facilities.

(e) Can be abandoned quickly in event of detection.

(f) Has an accessible ALTERNATE AREA.

(2) *Reconnaissance.* Upon arrival at clandestine assembly area, reconnoiter to insure the following:

(a) The area is free of aggressor forces.

(b) Sufficient natural cover and concealment is offered.

(c) Area is easy to defend.

(d) Early warning systems can be established.

(e) Provides covered routes to other assembly areas.

(f) Covered routes to and from alternate areas.

(g) Protection from weather and an adequate supply of water if area is to be occupied for a period of time.

(3) *Occupation.* When a unit moves into a clandestine assembly area it should—

(a) Establish an early warning system with communication to the main body.

(b) Establish security within the area.

(c) Select a command post and sectors of responsibility.

(d) Determine work priorities—

1. Individual protection.

2. Care and cleaning of weapons.

3. Redistribution of ammunition and equipment.

4. Personal hygiene.

5. Feeding.

6. Rest.

(e) Disposal holes must be dug for all items discarded and camouflaged when closed.

(f) Select an alternate area with both primary and alternate routes.

(g) Enforce movement, noise, light, and fire discipline.

(h) Restore area to natural appearance before departure.

5. Cliff Assault Techniques

a. Objective. To acquaint the Soldier with the techniques and methods of assaulting cliffs, beach landings, scaling techniques, beach and cliffhead security, and organization of the raiding party.

b. Reference. Paragraphs 42 through 45.

c. Requirements.

(1) *Cadre.* One principal instructor.

(2) *Training aids.* Blackboard, chalk, eraser.

d. Instructor's Notes (one hour).

(1) *Introduction.*

(2) *Demonstration and explanation of use of special equipment.*

(a) *Scaling ladders.*

(b) *Toggle ropes.*

(c) *Grappling hooks.*
(d) *Bear claws.*
(e) *Rockets with grappling heads.*
(f) *Mountaineering equipment* (hammer, piton, snap link, sling rope, etc.).

(3) *Organization and duties of cliff assault elements.*
(a) *Climbers.*
(b) *Security.*
(c) *Control.*
(d) *Main force.*
(e) *Cliffhead team.*

(4) *Evacuation of cliffhead.*

6. Combat Formations

a. Objective. To familiarize the Soldier with common arm-and-hand signals, the tactical advantages of combat formations, and the application of prescribed formations when subjected to enemy fire.

b. References. FM's 7–10, 21–6, and 21–75.

c. Requirements. Cadre: one principal instructor per platoon-size unit.

d. Instructor's Notes.

(1) Too often arm-and-hand signals are forgotten by leaders and/or patrol members tend to neglect watching for correct signals. Stress continuous use of and observation of signals for maximum control.

(2) Major factors influencing choice of formations are—
(a) Mission.
(b) Terrain.
(c) Visibility.
(d) Weather.
(e) Enemy situation.
(f) Control.
(g) Speed.
(h) Security.
(i) Dispersion.
(j) Stealth.

(3) Review types of formations and applicable situations.

(4) Practical work in signaling and movement from one formation to another.

7. Confidence Tests

a. Objective. To increase the confidence of the Soldier by requiring him to negotiate obstacles which appear more difficult than they actually are. The tests are conducted when the individual is tired in both mind and body, and when his self-confidence is at its lowest ebb.

b. Reference. TM 21-200.

c. Requirements.

(1) *Cadre.*

(*a*) One principal instructor.

(*b*) One assistant instructor (safety officer).

(*c*) One assistant instructor (demonstrator).

(*d*) Necessary medical support.

(2) *Training aids.*

(*a*) Diagram of confidence test.

(*b*) Confidence test.

d. Confidence Tests.

(1) Negotiation of a confidence course may be sufficiently strenuous to be an excellent physical conditioner.

(2) Should be negotiated at slow speed.

(3) Listed below and shown in figure 33 are some suggested types of confidence tests. The instructor should consider the safety aspects of any type test but should not limit himself to stereotype tests. Free climbs, water drops from helicopters or helicopters rappels can all be performed with acceptable safety standards.

(*a*) Ladder constructed of logs to height of 5 meters or more. Requirement: to climb.

(*b*) Rope stretched between two towers, trees, or similar anchor point at height of 5 meters or more. Requirement: to hand walk over an obstacle.

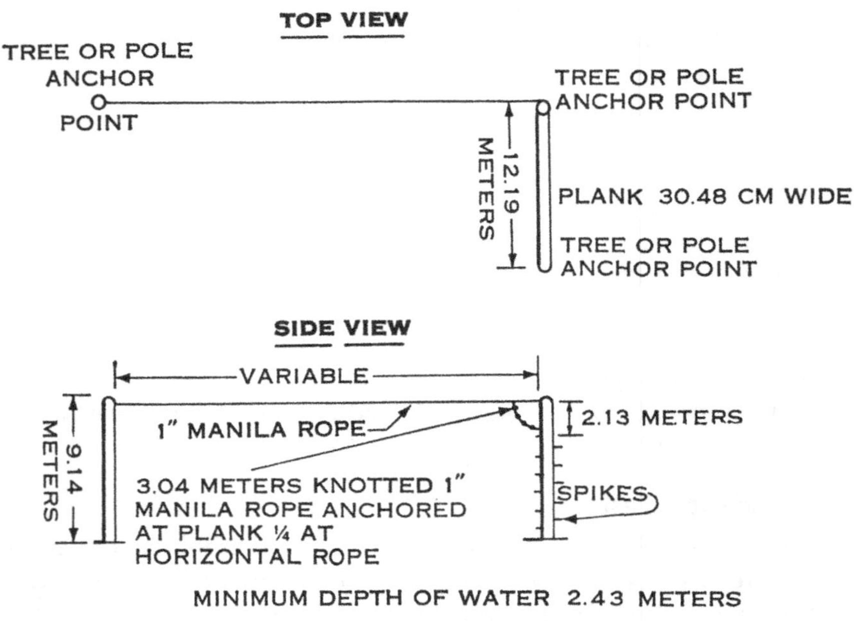

1 Diagram of rope drop confidence test

Figure 33. Confidence tests.

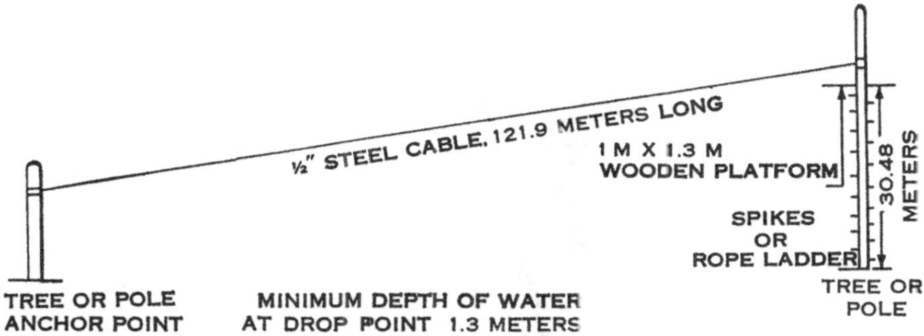

2 Diagram of suspension traverse confidence test

3 Confidence tests are easy to devise and contribute to an effective Ranger training program. Above log balance is 4.3 meters high. Soft dirt or sawdust cushions the Soldier's fall should he lose his balance.

Figure 33—Continued.

(*c*) Rope suspended from overhang. Requirement: to swing across an obstacle.

(*d*) Combination of ropes, ladders, or planks arranged at various heights over water at least 3 feet deep. Requirement: climb ladder, walk plank and hand climb rope before dropping into water from height of 12 meters or more.

(*e*) Slides requirement: to slide on a pulley or rope from heights of 7 meters or more over bodies of water.

4 Increased individual confidence results from overcoming seemingly difficult obstacles

Figure 33—Continued.

8. Current Enemy Situation

a. Objective. To maintain a combat atmosphere during the instruction by introducing the latest combat reports and developments of frontline and enemy units. For best results, situation reports should be integrated repeatedly throughout the instruction. These are based on a previously prepared scenario. The scenario presents missions which require combat and reconnaissance patrol action in the training program.

b. Requirements.

(1) *Cadre.* One principal instructor.

(2) *Training aids.* Map overlays.

c. References. Pertinent field manuals, staff officer's manual.

9. Demolitions

a. Objective. To acquaint the Soldier with the preparation, calculation, and placement of various military demolitions, charges, and explosives.

b. References. FM's 5–25, 5–31, and 20–32; TM's 9–1900 and 9–1946.

c. Requirements.

(1) *Cadre.*

(*a*) One principal instructor.

(*b*) One assistant instructor.

(*c*) Six lane instructors.

(*d*) Support. Two medical aidmen.

(2) *Vehicle.* One ambulance, one ammunition vehicle.

(3) *Equipment.* Podium, crimpers, matches, whistle (safety device), posthole digger, sandbags, shovels, friction tape, one table per ten students, one four-foot length Marline per two students, roll of Marline or heavy string, 49 meters of one-half or one-quarter inch rope, galvanometer, blasting machine, fire extinguishers, cover for stored explosives, material to be cut or destroyed.

(4) *Training aids.* Blackboard, eraser, chalk, dummy charges, ropes, detonating cord clips, pointer, dummy caps, priming adapters.

d. Instructor's Notes (1–3 hours basic).

(1) *Introduction.* Fire 2 one-pound charges of explosive to gain attention. Discuss military importance of all types of explosives and their uses.

(2) *Development* (FM 5–25).

(*a*) *Definition and classification of explosives.* (Examples of each and demonstrate.)

(*b*) *Desirable characteristics of military explosives.*

(*c*) *Types and characteristics.* Exhibit blocks, cap wells, plasticity, etc.

1. TNT and tetrytol. High explosives.

2. Composition C3 and C4. Plastic explosives.

3. Ammonium nitrate and nitrostarch.

4. Dynamite, blasting explosive.

(*d*) *Demolition equipment.* Demonstrate and discuss items needed to make nonelectric primers.

1. Blasting caps, nonelectric.

2. Detonating cord.

3. Cap crimpers.

4. Priming adapters and detonating cord clips.

5. Time fuze, safety fuze M700, and fuze lighters, matches, etc.

(*e*) *Practical work.* Using dummy items, students prepare simple primers and detonating cord firing systems. Correct crimping technique, knot tying, and use of cord clips are practiced in preparing a main and branch ring detonating cord firing system.

(*f*) *Summary and questions of class.*

e. Instructor's Notes (1–3 hours advance) (4–7 hours basic).

(1) *Introduction.*

(2) *Development.*

(*a*) Explanation and demonstration of the satchel charge, pole charge, bangalore torpedo, and shaped charge.

1. Satchel charge. Preparation, priming and use.

2. Pole charge. Preparation, priming, and use.

3. Bangalore torpedo. Description, preparation, priming, and use.

4. Shaped charge. Description, preparation, priming, and use.

(*b*) Upon completion of each subtopic above, the students will report to the proper lane instructor's worktable and prepare the appropriate charge. During the preparation, the lane instructor continues the instruction by asking questions. The entire group moves down range on order from the principal instructor. Blasting caps are issued down range and the charge is primed by two students. All students and instructors move to firing positions. On order from the principal instructor, the students fire the charge, examine the results of detonation, and continue with the instruction.

(*c*) Discussion of special purpose charges.

1. Timber cutting charges. Placement and calculation.

2. Steel cutting charges. Placement and calculation for I-beams, chains, rods, bars, channels, cables, structural and laminated steel.

3. Cratering charges. Placement and calculation.

4. Pressure and breaching charges. Placement and calculation for concrete bridges and abutments.

5. Special shape charges.

6. Relative effectiveness of explosives as it applies to charge calculations using the formula.

(*d*) Students receive demolition cards (fig. 34). Report to lane instructor and work sample problems involving calculation and placement of charges on various type installations, objects for destruction, or objects for cutting.

(*e*) If sufficient time is available for additional training, it is possible to cover a variety of demolition topics to include:

electric blasting caps, electric blasting machines, electrical firing systems, use of electric and nonelectric systems in boobytrap devices, types of boobytraps, and installations.

(*f*) Students report to lane instructor for preparation of timber and steel cutting charges. All students are required to compute the charge required for the type timber or steel to be cut. The correct solutions are used, the charge prepared and placed. Blasting caps are issued after the charge has been placed and the charges are primed. All students and instructors move to firing positions. The charges are detonated on order from the principal instructor.

(*g*) Summary and questions on the instruction.

f. Instructor's Notes (4–7 hours advance).

(1) Introduction.

(2) Development. Explanation, demonstration, and Soldier preparation of the following special charges and expedients.

(*a*) *Improvised shaped charge* (*Monroe*) (fig. 35).

1. Standoff—1½ times diameter of cone.

RELATIVE EFFECTIVENESS OF PRINCIPAL MILITARY EXPLOSIVES AS EXTERNAL CHARGES			SIZE ISSUED
1.00	TNT	(BLOCK)	1 LB
		(BLOCK)	1/2 LB
1.34	COMPOSITION C-2	(M-3 DEMOLITION BLOCK)	2-1/4 LB
		(M-4 DEMOLITION BLOCK)	1/2 LB
	COMPOSITION C-3	(M-3 DEMOLITION BLOCK) (M-4 DEMOLITION BLOCK) (M-5 DEMOLITION BLOCK)	2-1/4 LB
	COMPOSITION C-4	(M-5 DEMOLITION BLOCK)	2-1/2 LB
1.20	TETRYTOL	(M-1 DEMOLITION BLOCK) (M-2 DEMOLITION BLOCK)	2-1/2 LB
.42	AMMONIUM NITRATE	CRATERING CHARGE	40 LB
.92	MILITARY DYNAMITE		
.78	60% STRAIGHT OR GELATIN DYNAMITE (COMMERCIAL)		
.93	GUNCOTTON, WET (BRITISH)		
.48	GELIGNITE (BRITISH COMMERCIAL)		
.94	PICRIC ACID		

QUANTITIES OF EXPLOSIVE IN THESE FORMULAS AND TABLES ARE FOR TNT, FOR OTHER EXPLOSIVES, DIVIDE THE QUANTITY FOR TNT BY THE EFFECTIVENESS RATIO (FROM THE LEFT COLUMN, ABOVE)

DON'T HANDLE EXPLOSIVES UNLESS YOU KNOW HOW!

1 Military explosives

Figure 34. Demolition cards.

PRESSURE CHARGES

POUNDS = $3H^2T$

WHERE H = HEIGHT OF BEAM IN FEET (INCLUDING ROADWAY)

WHERE T THICKNESS OF BEAM IN FEET

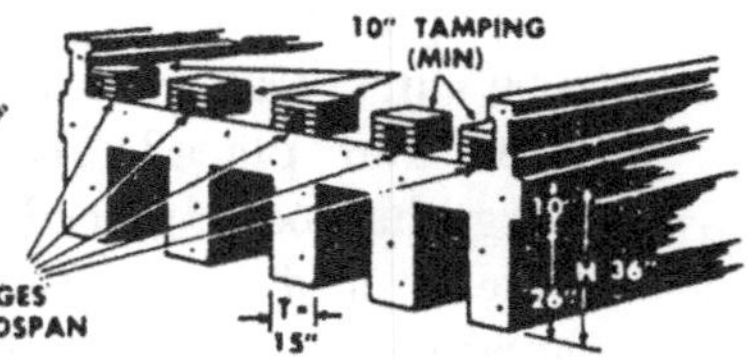

POUNDS OF EXPLOSIVE FOR EACH BEAM (TAMPED CHARGES)*									
HEIGHT OF BEAM IN FEET	THICKNESS OF BEAM IN FEET								
	1 (12 IN)	1-1/4 (15 IN)	1-1/2 (18 IN)	1-3/4 (21 IN)	2 (24 IN)	2-1/4 (27 IN)	2-1/2 (30 IN)	2-3/4 (33 IN)	3 (36 IN)
1 (12 IN)	3								
1-1/4 (15 IN)	5	6							
1-1/2 (18 IN)	7	9	11						
1-3/4 (21 IN)	10	12	14	16					
2 (24 IN)	12	15	18	21	24				
2-1/4 (27 IN)	16	19	23	27	31	35			
2-1/2 (30 IN)	19	24	29	33	38	43	47		
2-3/4 (33 IN)	23	29	34	40	46	51	57	63	
3 (36 IN)	27	34	41	48	54	61	68	75	81
3-1/4 (39 IN)	32	40	48	56	64	72	80	88	95
3-1/2 (42 IN)	37.	46	56	65	74	83	92	101	111
3-3/4 (45 IN)	43	53	64	74	85	95	106	116	127
4 (48 IN)	48	60	72	84	96	108	120	132	144
4-1/4 (51 IN)	55	68	82	95	109	122	136	149	163
4-1/2 (54 IN)	61	76	92	107	122	137	152	167	183
4-3/4 (57 IN)	68	85	102	119	136	153	170	187	203
5 (60 IN)	75	94	113	132	150	169	188	207	225

*INCREASE AMOUNTS BY 1/3 WHEN CHARGES ARE UNTAMPED

BRIDGE ABUTMENT DESTRUCTION

ABUTME TS 5 FEET OR LESS IN THICKNESS

BEGINNING 5 FEET IN FROM SIDE OF ROAD, PLACE 40 LB CRATERING CHARGES IN HOLES 5 FEET DEEP, 5 FEET ON CENTERS AND 5 FEET EHIND RIVER FACE OF THE ABUTMENT.

ABUTMENTS MORE THAN 5 FEET IN THICKNESS

CALCULATE CHARGES BY BREACHING FORMULA AND PLACE AGAINST REAR FACE AT A DEPTH E UAL TO THICKNESS OF ABUTMENT AND SPACE THE SAME AS OTHER BREACHING CHARGES.

(WHEN ABUTMENT IS OVER 20 FEET HIGH, ADD A ROW OF BREACHING CHARGES ON THE RIVER FACE AT THE BASE OF THE ABUTMENT)

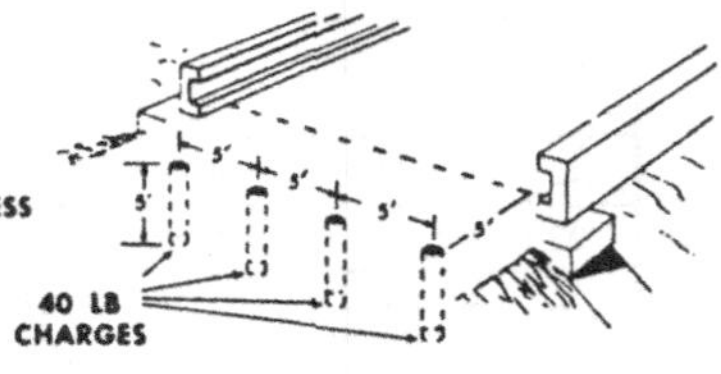

2 Pressure charges

Figure 34—Continued.

STEEL CUTTING CHARGES

COMMON STEEL MEMBERS

POUNDS = 3/8 x AREA OF CROSS SECTION IN SQ IN.

(CALCULATE RECTANGULAR AREAS AND ADD TO OBTAIN T TAL AREA)

EXAMPLE PROBLEM

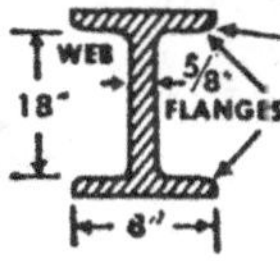

FLANGES: WIDTH 8 IN
THICKNESS = 1 IN
FROM TABLE, CHARGE 3.0

WEB: WIDTH 18 IN
THICKNESS 5/8 IN
FROM TABLE, CHARGE = 4.3

CHARGE: 2 FLANGES = 2 x 3.0 = 6.0
WEB = 1 x 4.3 = 4.3
T T L 10.3
11 POUNDS)

CABLES, RODS AND BARS

POUNDS . AREA OF CROSS-SECTION IN SQUARE INCHES WHEN DIAMETER OR DIMENSION IN CONTACT WITH EXPLOSIVE EQUALS 2 INCHES OR LESS (WHEN MORE USE P 3.8 A)

POUNDS OF EXPLOSIVE FOR RECTANGULAR STEEL SECTIONS OF GIVEN DIMENSIONS

AVERAGE THICKNESS OF SECTION IN INCHES	WIDTH OF SECTION IN INCHES												
	2	3	4	5	6	8	10	12	14	16	18	20	24
1/4	.2	.3	.4	.5	.6	.8	1.0	1.2	1.3	1.5	1.7	1.9	2.3
3/8	.3	.5	.6	.7	.9	1.2	1.4	1.7	2.0	2.3	2.6	2.8	3.4
1/2	.4	.6	.8	1.0	1.2	1.5	1.9	2.3	2.7	3.0	3.4	3.8	4.5
5/8	.5	.7	1.0	1.2	1.4	1.9	2.4	2.9	3.3	3.8	4.3	4.7	5.7
3/4	.6	.9	1.2	1.4	1.7	2.3	2.8	3.4	4.0	4.5	5.1	5.7	6.8
7/8	.7	1.0	1.4	1.7	2.0	2.7	3.3	4.0	4.6	5.3	6.0	6 6	7.9
1	.8	1.2	1.5	1.9	2.3	3.0	3.8	4.5	5.3	6.0	6.8	7.5	9.0

TO USE TABLE:

1. MEASURE RECTANGULAR SECTIONS OF MEMBER SEPARATELY.
2. USING TABLE, FIND CHARGE FOR EACH SECTION.
3. ADD CHARGES FOR SECTIONS TO FIND TOTAL CHARGE.
4. NEVER USE LESS THAN CALCULATED CHARGE.

3 Steel cutting charges

Figure 34—Continued.

CRATERING CHARGES

DELIBERATE ROAD CRATER

ALTERNATE 5-FOOT AND 7-FOOT HOLES SPACED AT 5-FOOT INTERVALS. (END HOLES ALWAYS 7-FOOT)

USE 40-LB CHARGES IN 5-FOOT HOLES AND 80-LB CHARGES IN 7-FOOT HOLES.
RESULTING CRATER APPROX. 8-FEET DEEP AND 25-FEET WIDE.

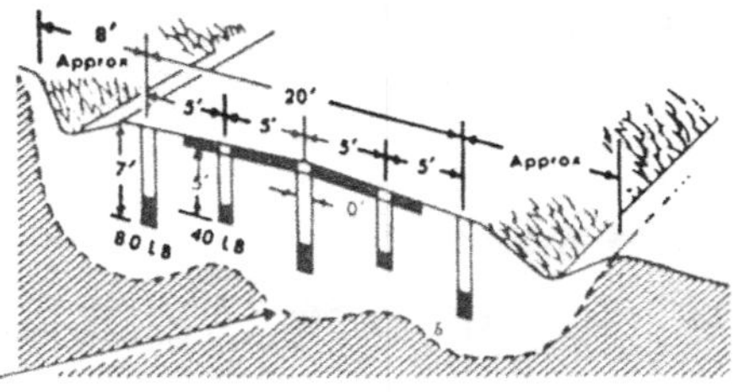

HASTY ROAD CRATER

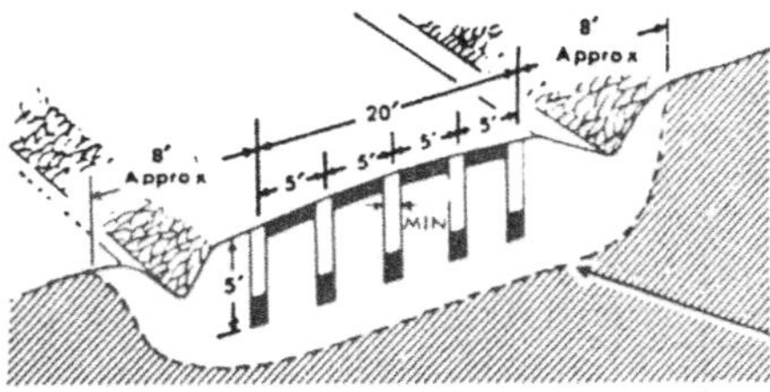

HOLES OF EQUAL DEPTH, (2½ FEET TO 5 FEET) SPACED AT 5-FOOT INTERVALS. USE 10-POUNDS OF EXPLOSIVES PER FOOT OF DEPTH. RESULTING CRATER DEPTH APPROX. 1½ TIMES DEPTH OF BOREHOLES. WIDTH APPROX. 5 TIMES DEPTH OF BOREHOLES.

TIMBER CUTTING CHARGES

INTERNAL CHARGES

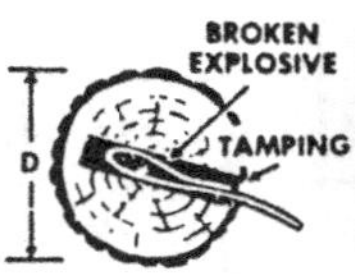

$$\text{POUNDS} = \frac{D^2}{250}$$

WHERE D IS THE LEAST DIMENSION IN INCHES

EXTERNAL CHARGES

$$\text{POUNDS} = \frac{D^2}{40}$$

WHERE D IS THE LEAST DIMENSION IN INCHES

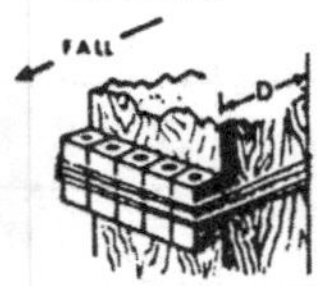

PLACEMENT	LEAST DIMENSION OF TIMBER IN INCHES									
	6	8	10	12	15	18	21	24	30	36
	POUNDS OF EXPLOSIVE									
INTERNAL	1/5	1/4	2/5	3/5	1	1-1/3	1-3/4	2-1/2	4	5
EXTERNAL	1	2	3	4	6	8	11	15	23	33

4 Cratering and timber cutting charges

Figure 34—Continued.

BREACHING CHARGES

POUNDS = R^3KC (ADD 10% IF LESS THAN 50 POUNDS)

R = THICKNESS OF MATERIAL IN FEET FOR EXTERNAL CHARGES

R = ½ THICKNESS OF MATERIAL IN FEET FOR INTERNAL CHARGES

K = MATERIAL FACTOR

C = TAMPING FACTOR (CHARGES SPACED AT 2 x R FEET APART)

POUNDS OF EXPLOSIVE FOR BREACHING REINFORCED CONCRETE

THICKNESS OF CONCRETE IN FEET	CHARGE IN POUNDS				DISTANCE BETWEEN CHARGES IN FEET	
	EXTERNAL			INTERNAL	EXTERNAL	INTERNAL
2	16	22	28	1	4	2
2-1/2	31	43	55	2	5	2-1/2
3	41	52	67	4	6	3
3-1/2	59	83	106	6	7	3-1/2
4	88	124	159	8	8	4
4-1/2	126	176	226	11	9	4-1/2
5	157	219	282	16	10	5
5-1/2	208	292	375	20	11	5-1/2
6	270	378	486	21	12	6
6-1/2	344	481	618	26	13	6-1/2
7	370	517	663	33	14	7
7-1/2	454	635	817	40	15	7-1/2
8	551	771	991	49	16	8

SHAPED CHARGES

	REINFORCED CONCRETE			ARMOR PLATE	
	DEPTH OF PENETRATION IN THICK WALLS	DIAMETER OF HOLE	DEPTH OF HOLE WITH SECOND CHARGE PLACED OVER FIRST HOLE	PERFORATION	AVERAGE DIAMETER OF HOLE
M3 SHAPED CHARGE	60	3-1 2	84	20	2-1 2
M2A3 SHAPED CHARGE	30	2-3 4	84	20	1-1 2

VALUE OF K

MATERIAL	R	K
EARTH	ALL VALUES	0.05
POOR MASONRY, SHALE AND HARDPAN, GOOD TIMBER, AND EARTH CONSTRUCTION	ALL VALUES	0.23
GOOD MASONRY, ORDINARY CONCRETE, ROCK	LESS THAN 3 FT	0.35
	3 FEET TO LESS THAN 5 FEET	0.28
	5 FEET TO LESS THAN 7 FEET	0.25
	7 FEET OR MORE	0.23
DENSE CONCRETE, FIRST-CLASS MASONRY	LESS THAN 3 FT	0.45
	3 FEET TO LESS THAN 5 FEET	0.38
	5 FEET TO LESS THAN 7 FEET	0.33
	7 FEET OR MORE	0.28
REINFORCED CONCRETE (CONCRETE ONLY; WILL NOT CUT REINFORCING STEEL)	LESS THAN 3 FT	0.70
	3 FEET TO LESS THAN 5 FEET	0.55
	5 FEET TO LESS THAN 7 FEET	0.50
	7 FEET OR MORE	0.43

VALUES OF C

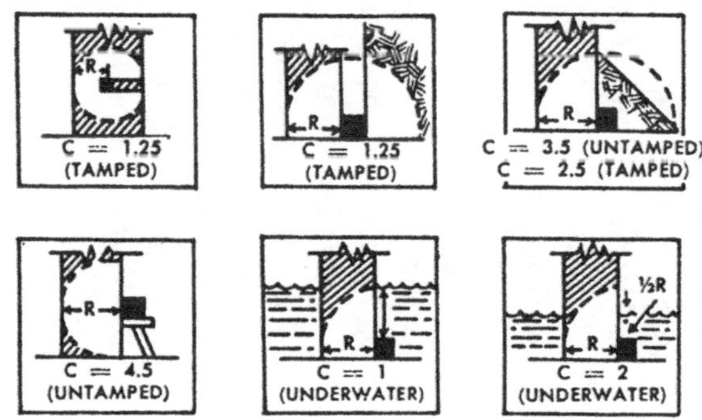

5 Breaching charges

Figure 34—Continued.

2. Size—2 times height of cone.
3. Angle—45 to 60 degrees.
4. Detonation—exact rear center of charge.

(*b*) *Diamond charge* (fig. 35).
1. Long axis—equal to circumference of target.
2. Short axis—equal to ½ circumference of target.
3. Depth—¼-inch for mild steel; ¾-inch for high carbon steel.
4. Detonation—simultaneous at each end of short axis.
5. Long axis wrapped completely around target.

(*c*) *Counterforce.*
1. Size—1 to1½ lbs per foot of concrete.
2. Placement—both charges exactly opposite each other and flush with target.
3. Detonation—simultaneous detonation in exact rear center of each charge.

(*d*) *Ribbon charge.*
1. Depth—½ thickness of target plus ⅛-inch.
2. Width—2 times thickness of target.
3. Length—same as length of cut desired.
4. Detonation—from one end only.

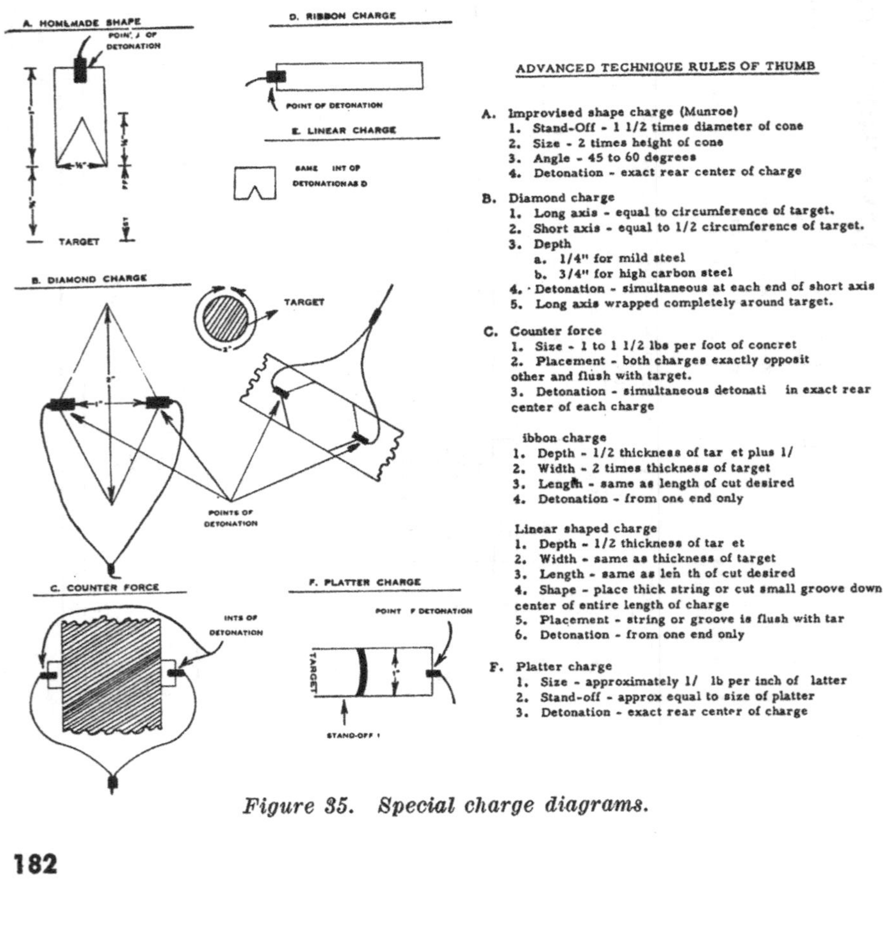

Figure 35. Special charge diagrams.

(*e*) *Linear shaped charge.*

1. Depth—½ thickness of target.

2. Width—same as thickness of target.

3. Length—same as length of cut desired.

4. Shape—place thick string or cut small groove down center of entire length of charge.

5. Placement—string or groove is flush with target.

6. Detonation—from one end only.

(*f*) *Platter charge.*

1. Size—approximately ½ lb per inch of platter.

2. Standoff—approximately equal to size of platter.

3. Detonation—exact rear center of charge.

(*g*) *Makeshift hand grenade.*

1. Select type and size desired.

2. Determine time delay desired (fuze burning rate).

3. Prime charge.

4. Attach bolts, nails, etc., for fragmentation.

(*h*) *Grape shot.*

1. Select container (#10 can).

2. Place inverted funnel in bottom of can filled with pieces of metal.

3. Fill container with explosives.

4. Prime exact rear center.

5. Implace reinforcing outside of container with earth.

(*i*) *Martini glass-shaped charge.*

1. Obtain a glass similar in shape to martini glass.

2. Break stem and use drinking portion in same fashion as funnel in preparing homemade shaped charge.

(*j*) *Molotov cocktail.*

1. Select quart or larger bottle and fill with ½ gas and ½ oil.

2. Secure cotton or rags to bottom of bottle.

3. Light rags and throw in tank engine vent.

(*k*) *Grenade necklace.*

1. Remove firing mechanism.

2. Connect at 3.5 meter intervals by detonating cord and twine.

3. Detonate electrically.

4. Used in ambush killing zone and ambush areas difficult to cover by fire.

(*l*) *Summary and questions on subject matter.*

10. Escape and Evasion

a. Objective. To acquaint the Soldier with methods of operation used by individual Soldiers in enemy territory. Includes planning, reconnaissance, evasion tactics, conduct if captured, and methods of escape.

b. References. DA Pam's 21–71, 30–101, 108–1, 355–6, 355–51; AR's 350–30, 350–225; TF 21–2720; and FM's 21–75, 21–77A.

c. Requirements.

(1) *Cadre.* One principal instructor, one assistant instructor.

(2) *Equipment.* Projector, screen, power source.

d. Instructor's Notes.

(1) When the situation dictates that evasion tactics be employed, a unit is organized into small groups.

(2) The senior officer or enlisted man present should take command of the group or unit and organize it for evasion tactics. He should utilize his subordinate leaders to insure effective leadership for the release.

(3) The senior in command will determine the release location(s) and times of release of groups. In addition, he will supervise the releases.

(4) To accomplish their mission, evaders must employ the principles they have learned from instruction in map reading, patrolling, individual combat techniques, evasion, escape, and survival.

(5) If captured, an evader must keep alert for an opportunity to escape as soon as possible. The longer he waits, the more difficult escape becomes.

(6) Upon being interrogated by the enemy, a member of the United States forces will give only his name, rank, serial number, and date of birth.

(7) Evaders who have infiltrated through enemy lines should seek a quiet sector in the friendly lines to make contact with friendly units. Utmost care must be exercised at this time.

Figure 36. Practical application in demolitions should be stressed.

(8) Evaders who make contact with friendly forces can give valuable tactical information and should be debriefed immediately.

11. Hand-to-Hand Combat

a. Objective. To familiarize the Soldier with hand-to-hand combat, develop self-confidence, improve physical fitness, and instill a spirit of aggressiveness and a will to win.

b. References. TM 21–200 and FM's 21–150 and 23–25.

c. Requirements.

(1) *Cadre.*

(*a*) One principal instructor.

(*b*) Five assistant instructors per 100 Soldiers.

(*c*) Three demonstrators (also assistant instructors).

(*d*) Support.

1. One sound equipment operator.

2. Two aidmen.

(2) *Vehicle.* One ambulance.

(3) *Training aids.*

(*a*) Sawdust pit or plowed area.

(*b*) PT platform.

(4) *Communications equipment.* One sound equipment set with four speakers.

d. Instructor's Notes.

(1) *Introduction.*

(*a*) Soldiers progress at different rates during hand-to-hand combat instruction. Hourly instructional outlines are intended as a guide only.

(*b*) Soldiers receiving instruction are paired off by size into buddy teams. Each new exercise is demonstrated by the phases and then once at normal speed. Each Soldier is then talked through the exercise by the phases one or more times prior to performing the exercise at normal speed. Special attention must be given to insure that each individual assumes the proper fall positions.

(*c*) Warm-up exercises are conducted at the beginning of each hour of instruction. This should include a review of the fall positions.

(*d*) Breakdown of instructional hours—

First Hour

1. Introduction (reasons for hand-to-hand combat).

(a) Excellent *physical conditioner and body toughener.*

(b) Builds spirit of *aggressiveness and instills will to fight.*

(c) Instills *confidence.*

(d) Provides *know-how* for setting up similar type training programs.

2. Fundamentals.

(a) Momentum (overhead throw).

Note. Demonstrators are used to emphasize fundamentals.

(b) Maximum strength against weakest point (wrist takedown).

(c) Use of any and all available weapons (kick opponent to the ground).

(d) Balanced position (on-guard position).

(e) Mental position (use of growl).

(f) Accuracy and speed.

3. Throws (those to be learned by the Soldier).

4. Takedowns (those to be taught).

5. Holds and counters (those to be taught).

(a) Knife attacks.

(b) Bayonet attacks.

6. Techniques for disarming sentry—demonstration (modified rear strangle takedown).

7. Pistol disarming—demonstration.

8. Conclusion of demonstration.

9. Warm-up exercises (TM 21–200).

10. On-guard position (FM 21–150).

11. Left and right side fall position (FM 21–150).

12. Hip throw (FM 21–150).

Second Hour

1. Overhead fall position (FM 21–150).

2. Front fall position. This is performed by leaning forward, keeping the body straight, until the center of gravity forces the man's body to the ground. The shock of the fall is absorbed by the hands, forearms, and toes. No other part of the body touches the ground.

3. Overshoulder throw (FM 21–150).

4. Overhead throw.

Phase (*1*). Number two man charges his buddy who grabs number two man's shoulders and takes one or more steps backward, maintaining his opponent's momentum.

Phase (*2*). Number one man drops to ground, placing leg into opponent's stomach area.

Phase (*3*). Number one man straightens leg and throws number two man over his head; maintains his grip on the opponent's shoulders, rights himself and strikes number two man, following up the attack.

1. Introduction to knife disarming (pars. 94–100, FM 21–150).
2. Vulnerable points (pars. 45 and 46, FM 21–150).
3. Counter to downward stroke (par. 94, FM 21–150).
4. Counter against the backhand slash (par. 99, FM 21–150).
5. Counter to sidearm slash.

   Phase (1). Step into your opponent with the left foot and block the blow with the left forearm, elbow up.

   Phase (2). Pivot 180° on the left foot, sliding the left forearm down your opponent's knife arm, grabbing the

Figure 37. ***Ranger training can be incorporated into any military organizational schedule. Participants shown above are United States Military Academy Cadets undergoing the one-week training program outlined in this manual.***

   wrist. At the same time, reach up with the right hand and grab your opponent's clothing. Your opponent's right arm is straight and over your right shoulder, your knees are bent and your opponent is pulled in close to your buttocks.

   Phase (3). Pull down with both hands, straighten your knee, and simultaneously throw your opponent over your shoulder.

Fourth Hour

1. Review first and second hours.

2. Rear takedown.

Phase (1). Grab your opponent by his shoulders from the rear and force him off balance; shift your weight to your left foot.

Phase (2). Pull back on your opponent's shoulders, simultaneously kicking him in the rear of the knees, knocking him to the ground.

3. Rear strangle takedown.

Phase (1). Punch your opponent in the left kidney and at the same time strike him in the Adam's apple with your right forearm, bending him to the rear.

Phase (2). Keeping your left fist in the kidney area, grasp your opponent's left shoulder with the right hand and tighten the strangle hold on the neck. At the same time move backwards, walking your opponent to the ground.

Phase (3). Tighten your strangle hold until your buddy signals that your strangle hold is effective. Lock your right hand to your opponent's shoulders with your chin.

4. Cross hock takedown.

Phase (1). Step into your opponent with your left foot, placing it slightly to the rear and left of your opponent's right foot. At the same time strike your opponent with both hands on his shoulders, bending him to the rear, grasping his clothing.

Phase (2). Forcefully bring your right leg through and to your opponent's rear in such a manner as to strike his legs from under him.

Fifth Hour

1. Review third hour.

2. Second counter to the downward stroke (FM 21–150).

3. Counter against the upward stroke (FM 21–150).

4. Second counter against the upward stroke (FM 21–150).

5. Alternate method used to channel your opponent's knife threat (cross right arm over left to form a face aim view).

Sixth Hour

1. Introduction to bayonet disarming.

2. Parry right (FM 21–150).

3. Counter to the short thrust (FM 21–150).

Figure 38. Hand-to-hand combat provides the Soldier with another means to fight or defend himself when he is unarmed.

4. Counter to the long thrust.

Phase (1). Parry opponent's thrust to the right, grabbing the rifle with both hands once it is diagonally across his body.

Phase (2). Execute phase two of the overhead throw.

Phase (3). Execute phase three of the overhead throw, pulling the rifle from your opponent.

5. Second counter to the short thrust.

Phase (1). Parry opponent's thrust to the right, at the same time crow-hopping to the left, facing the right side of the rifle, both hands above the weapon.

Phase (2). Drop both hands, slamming the bayonet into the ground.

Phase (3). Grasp the rifle butt with the left hand and your opponent's right shoulder with the right hand, throwing your opponent over the rifle. Remove the rifle from the ground and finish off your opponent.

Seventh Hour

1. Review all unarmed hand-to-hand combat.

2. Execution of hold and counters to holds (FM 21–150).

3. Mass competition. Have each man attempt to throw the winners of other teams. The last person standing in the pit is the class champion.

12. Inspections

a. Objective. To require the Soldier to maintain a high standard in personal appearance and maintenance of equipment.

b. Scope. Inspections, scheduled in 30-minute periods, are designed to affect the tendency of tired men to neglect personal hygiene, clothing, and equipment. The Soldiers are inspected by the most senior and experienced officers and NCO's. A demerit system is established, in addition reinspections or additional training is given to those receiving an excessive amount of demerits.

(1) Inspections are part of Ranger training.

(2) Through frequent inspections, the Soldier is continually confronted with an exacting situation. Eventually, the mental confusion and physical indecision are replaced by calm mental confidence and physical dexterity in successful preparation for inspections.

(3) In addition to mental adjustment to seemingly harassing requirements, the Soldier acquires a higher standard of appearance and more orderly personal habits.

(4) The Soldier acquires a new respect for his clothing, equipment, and weapons.

13. Intelligence

a. Objective. To familiarize the individual Soldier with the importance of intelligence by showing him his responsibilities and in the collecting, recording, and forwarding of information. Secondly, to familiarize him with the procedures for processing enemy personnel, documents, and equipment.

b. References. FM's 21–75 and 30–7, and DA Pam 21–81.

c. Requirements.

(1) *Cadre.* One principal instructor.

(2) *Training aids.* Blackboard, chalk, eraser.

d. Instructor's Notes.

(1) Ranger training requires additional stress on the importance of combat information. Reconnaissance patrols, prisoner raids, etc., provide the basis for many Ranger type missions.

(2) Combat intelligence training should be scheduled prior to scheduling reconnaissance patrols.

(3) Prisoner handling is a primary task of Ranger trained personnel.

(4) Use of binoculars, sniperscope, and other infrared devices aid night observation and reconnaissance.

14. Map and Aerial Photograph Reading

a. Objective. To provide a review of the basic principles and techniques of map and aerial photograph reading, which the small unit leader can use and apply in combat. To develop in the individual a working knowledge of the basic principles of day and night field navigation; through application develop confidence in his abilities.

b. References. FM's 21–26 and 21–31.

c. Requirements.

(1) *Cadre.*

(*a*) One principal instructor, two assistant instructors.

(*b*) Support. One aidman, one light truck driver per 2½-ton truck.

(2) *Vehicles.* One ambulance, one 2½-ton truck per twenty Soldiers.

(3) *Equipment.* Twenty-five (3.81 cm x 7.62 cm x 1.22 m) stakes, 25 empty caliber .30 MLB ammunition cans, 25 flashlights.

(4) *Training aids.*

(*a*) Blackboard, chalk, eraser.

(*b*) Maps (at least one per buddy team).

(*c*) Pace course markers.

(*d*) Compass aiming stakes.

d. Instructor's Notes.

(1) *Introduction.*

(*a*) Combat reports of World War II and Korea referred to small unit operations that failed because of the inability of unit leaders to navigate in any type terrain, with or without accurate maps. For the most part, German operations conducted along the Russian front were conducted without any map coverage at the company level; yet these operations were successful initially, due to the average Soldier's navigational ability.

(*b*) Ranger operations are most often conducted at night and over the most difficult terrain available. Small combat operations were successful initially, due to the average surprise; to strike where least expected. To move over routes too difficult for the enemy to occupy requires that one must be an expert in terrain navigation.

(2) *Basic* (1–5 hours).

(*a*) Conference map reading to include marginal information, topographical and military symbols, military grid coordinates, map orientation scale, direction, azimuth, declination diagram, G-M angle, relief and topography, profiles, intersection and resection, types of aerial photos, including basic photo interpretation, point designation, grids, scale, and method of determining magnetic North.

(*b*) Practical application on each subject listed above.

(3) *Test* (1 hour). This test should be designed to include only realistic questions applicable to combat situations.

Examples.

1. At coordinates DR 66105016 there is an observer with a compass but no map. You are in contact with him by telephone and you want him to keep the building at DR 66204925 under observation. What instructions would you give him?

Answer:____

2. From your position, you can see a railroad and road junction that you know to be KENWOOD in grid square 6547 at a magnetic azimuth of 175°. You can also see the northeast end of OUTLAW landing strip in grid square 6352 on a magnetic azimuth of 307°. Describe your position:

____

3. From your position on HILL 552, coordinates 641527, you have been given a mission to run a patrol to the buildings at coordinates 55643970. What magnetic azimuth will you go on?__________________ What is the distance in meters?__________________________ Estimated time in hours and minutes it will take your six-man patrol to move from your position to the objective:_____________

(4) *Use of map and compass* (1 hour). Conference on methods of determining north by use of a watch; use of the sun and stars; for navigation; use of the compass (to include methods used for day and night readings); use of the offset during navigation; use of the pace; and use of terrain features in night and day movement. The class ends with the Soldier checking his compass by using compass stakes and walking a 200-meter pace course to determine his average pace.

(5) *Night compass course.* Principal instructor briefs Soldier on the outline of compass course to include the use of control roads for medical evacuation. Soldiers are moved to the starting stake along the control road. One assistant instructor patrols the control roads and picks up buddy teams which become lost on the course and returns them to their last correct stake.

Discussion—night compass course.

1. Once an area has been selected, easily identified terrain features are located on the map. Using these as base locations, tentative stake positions are selected throughout the course and rough routes are selected. Direction cards, using every method of navigation, are made out. The directions start each buddy team along a control road or starting stake and guide them to eight numbered stakes. The routes are selected so that the ninth stake is the starting stake.

2. Distances between the stakes vary from 400–1200 meters. Different routes are set up permitting the use of each stake more than one time. The stakes are numbered so that the students may record them on a mimeographed handout.
3. When each instruction card has been checked by more than one person for accuracy, the stakes are implanted. The routes should be checked for accuracy on the ground; however, if the map is correct and the stakes are placed using extreme care, the routes should be accurate.
4. Example instructional cards, figure 39.

BLUE ROUTE

PROCEED FROM THIS POINT ON A MAGNETIC AZIMUTH OF 300 DEGREES UNTIL YOU REACH AN INTERMITTENT STREAM BED. PROCEED DOWN STREAM FOR 300 METERS UNTIL YOU REACH A TRAIL RUNNING EAST-WEST. PROCEED ON AN AZIMUTH OF 240° FOR 600 METERS TO REACH YOUR NEXT STAKE.

GREEN ROUTE

PROCEED WEST 400 METERS FROM THIS POINT UNTIL YOU REACH A HILL TOP. FROM THIS POINT PROCEED ON AN AZIMUTH OF 60° FOR A DISTANCE OF 500 METERS TO REACH YOUR NEXT STAKE.

BLUE ROUTE

USING THE MAP ATTACHED TO THIS CARD, LOCATE YOUR PRESENT POSITION AND PROCEED BY ANY ROUTE TO GRID COORDINATES 4576 0245 TO REACH YOUR NEXT STAKE.

Figure 39. Example instructional cards.

15. Mountain Techniques and Expedients

a. Objective. To familiarize the Soldier with basic military mountaineering techniques, equipment, and expedients.

b. References. FM 31–72; TF's 7–1475, 7–1480, 7–1518; and appendix V.

c. Requirements.

(1) *Cadre.* Five per 100 Soldiers.

(2) *Equipment.* For nomenclature and stock number of mountaineering equipment, see chapter 3.

d. Instructor's Notes (1–2 hours).

(1) Orientation on mountain climbing and display of climbing equipment.

(2) Demonstration of British crawl and sling seat (Tyrolean traverse) method.

(3) Six basic knots and rappel seat.

(*a*) Square knot.

(*b*) Double sheet bend.

(*c*) Round turn and two half hitches.

(*d*) Clove hitch (both methods).

(*e*) Bowline knot.

(*f*) Butterfly knot.

e. Instructor's Notes (3–9 hours).

(1) Rope installation. Anchors and knots.

(2) Rope bridges. One, two, and three ropes and method of crossing.

(3) Suspension traverse.

(4) Rappelling.

(*a*) Body rappel.

(*b*) Hasty rappel.

(*c*) Seat rappel (hip method).

(5) Balance climbing and mountain walking.

(6) Application and practical work to complete remaining hours.

f. Guide. This instruction is best given to small sections of the training group with these sections rotating at periodic intervals.

16. Orientation

a. Objective. To familiarize the Soldier with the training to be conducted, standards to be maintained, training procedures, the aggressor enemy, and the existing combat situation.

b. Scope. The orientation is scheduled immediately after the training unit arrives in the bivouac or training area and is presented by the cadre training officer. The orientation is given in an operation order format and presents only administrative and operational matters which could logically support the combat situation.

17. Patrol Planning, Orders, and Techniques

a. Objective. To instill in the Soldier the proper basic techniques of planning, preparing, and executing any Ranger patrols or missions.

b. References. FM's 7–15, and 21–75; TF's 7–1714, and 7–1750.

c. Requirements.

(1) *Cadre.* One principal instructor, one assistant instructor.

(2) *Support.* One demonstrator in patrol uniform, projector operator.

(3) *Equipment.* Blackboard, chalk, eraser, projector set, patrol uniform, blowup type training aids of patrol leader's order, patrol report form.

d. Instructor's Notes (1–4 hours).

(1) *Introduction.* Patrolling operations are of paramount importance in the conduct of combat operations and collection of combat information. The success of pending operations by larger units often depends entirely on the successful operations of patrols and patrol-size units.

(*a*) *Classification.* Patrols are classified according to the mission they perform. The two general classifications are combat and reconnaissance patrols.

1. Reconnaissance. Gathers information pertinent to assigned mission. Normally a five- to nine-man patrol is broken down into two elements, reconnaissance and se-

Figure 40. Body and seat rappels are best suited for Ranger type operations.

Figure 41. The seat rappel is an excellent confidence builder.

curity groups. The reconnaissance element, which observes and gathers information, is composed of two or three men. The security element which provides protection for the reconnaissance element is composed of the remaining members, normally three to six men. Organization of the patrol varies according to the mission.

2. *Combat.* Destroys or captures enemy personnel, equipment, installations, or provides security. Organized into two basic elements, a security element and an assault element. A third element, the fire support element, is formed when the mission requires a large fire supporting force. Normally, the fire support element maintains a

team status and is organic to the assault element. The security element is responsible for the security of the patrol at the objective area. The assault element closes with and destroys or captures the enemy and his equipment. The assault element may include such elements as a prisoner team, demolitions team, a special weapons team, etc. Upon completion of the mission, the assault element withdraws to the rally point under the protection of the support team.

(*b*) *Mission.* Patrol missions are based on requirements of higher headquarters and/or the impending tactical plan. Only one primary mission is assigned to each patrol.

(*c*) *Training.* Successful patrol preparation and conduct relies almost entirely on individual tactical training and its application to team performance to insure efficient patrolling. Subjects outlined in this appendix, applicable to the individual, must be mastered, as well as other phases of individual combat training outlined in FM 7–15. Efficient patrolling techniques insure maximum opportunity to accomplish the mission and minimum risk to the patrol.

(2) *Planning and patrol.* Daily patrol plans are prepared by higher headquarters. The latest maps, aerial photographs, and sketches used to plan missions are made available to the patrols.

(*a*) *Upon receipt of the mission.* The patrol receives its mission as part of an operations order. This order is given to the patrol leader and selected patrol members at the CP or other vantage point. The order may be given by the commander and/or his staff (S2, S3). Normally the S2 plans reconnaissance patrols and the S3 plans combat patrols; close coordination is necessary between these staff officers. The patrol briefing should include but not be limited to—

1. The current enemy situation in all available details.

2. All available information on the terrain.

3. The location of known or suspected enemy ambush points.

4. The location of known or suspected enemy strong points.

5. Best known crossing sites of rivers, valleys, mountains, etc.

6. Detailed mission of the patrol.

7. Information on any other patrols operating in the area.

8. Location of bombline, when and where applicable.

9. Plan for ground and air support.

10. Fire support available.

11. Communications available.

12. Current challenge and password.

13. Place and time of debriefing. A debriefing is not designed to be an operations order; however, it may follow that format if specific missions are assigned to subordinate units.

(*b*) *Tentative plan.* Based on the information received in the briefing, the patrol leader formulates a tentative plan. This plan should include but is not limited to:

1. Terrain study from map, photo, or sketch.

2. Quick estimate of the situation.

3. Arrange movement of unit.

4. Select reconnaissance route, schedule, personnel.

5. Plan warning order.

(*c*) *Warning order.* Issue warning order based on tentative planning to include time and place patrol leader's order will be issued. Use patrol warning order card as a guide (fig. 42).

(*d*) *Reconnaissance.* Physically examine the terrain to be used by the patrol according to (*b*)*1* above. Modify the preliminary plan as necessary and make final coordination with units concerned with the patrol.

The warning order should be issued to all members of the patrol by the patrol leader; however, there may be times when the patrol leader will use the chain of command to issue a patrol warning order. The patrol warning order should consist of the following minimum items of information.

a. A brief statement of the enemy and friendly situation.

b. Mission of the patrol.

c. Personal uniform, personal equipment to be carried by each patrol member, identification, camouflage measures (individual and equipment) and weapon each member will carry.

d. The complete chain of command (designate one man to supervise patrol members' preparation during patrol leader's absence).

e. Specific instructions to individual patrol members when and where to obtain necessary rations, water, ammunition and any other special equipment required.

f. Set a time and place for the patrol to receive the patrol order.

Figure 42. Patrol warning order.

(*e*) *Prepare and issue patrol order.* Upon completion of an estimate of the situation and a final plan, the patrol order should be prepared according to format (fig. 43).

(3) *Preparing the patrol.* The patrol uniform is prescribed by the patrol leader. Normally, it will consist of fatigues, soft cap, cartridge or pistol belt, canteen, first aid packet and pouch, bayonet with scabbard, poncho, boots, utility rope, ID tags,

1. SITUATION
   a. Enemy Forces: Weather, terrain, identification, location, activity, strength.
   b. Friendly Forces: Mission of next higher unit, location and planned actions of units on right and left, fire support available for patrol, mission and routes of other patrols.
   c. Attachments and Detachments.
2. MISSION - What the patrol is going to accomplish.
3. EXECUTION - (Sub-paragraph for each subordinate unit.)
   a. Concept of operation.
   b. Specific duties of each individual or unit.
   c. Coordinating instructions.
      (1) Time of departure and return
      (2) Formation and order of movement
      (3) Route and alternate route
      (4) Passage of friendly positions
      (5) Rallying points and actions at rallying points
      (6) Actions on enemy contact
      (7) Actions at danger areas
      (8) Actions at objective
      (9) Rehearsal
      (10) Reporting results and debriefing
4. ADMINISTRATION AND LOGISTICS
   a. Rations.
   b. Arms and ammunition.
   c. Special equipment (state which members will carry and use) and uniform.
   d. Method of handling wounded and prisoners.
5. COMMAND AND SIGNAL
   a. Signal.
      (1) Signals to be used for control within the patrol.
      (2) Communication with higher headquarters - radio call signs, times to report and special code to be used.
      (3) Challenge and password and code words.
   b. Command.
      (1) Chain of command.
      (2) Location of patrol leader and assistant patrol leader in formation.

Figure 43. Patrol order.

and weapon. Pack suspenders or harness is worn to aid in carrying back or shoulder loads; securing other equipment; and serving other utilitarian purposes. For night operations, the uniform may be altered and additional equipment carried.

(*a*) *Day fighter.* Top button of fatigues is buttoned. Steel helmet, if worn, is covered with camouflage material to break the outline. Soft cap is preferred. Weapons are dulled with soot or dirt; slings are taped or removed; stock is muffled and stacking swivel taped to the barrel. Poncho is folded over the pistol belt in the rear. Face, arms, neck, hands, etc., are camouflaged with grease paint or field expedient.

(*b*) *Night fighter*. Similar to day fighter except:

1. Sneakers may replace boots (short distance patrols).

2. Compass, used as signaling device as well as direction finding, is secured around the neck with boot lace. This prevents fumbling in the dark and loss of the item.

3. Carry flashlight with filter. A poncho makes an effective shelter for use with a light.

(*c*) *Equipment*. Patrols will be augmented when possible with special equipment, special weapons, and special transportation according to the dictates of weather, terrain, and missions assigned. Normal equipment lists to be checked by patrols are included in paragraph 26. Special items could include—

1. Scout dogs.

2. Helicopters, United States Army aircraft or United States Air Force aircraft.

3. Amphibious craft.

4. Submarine, pneumatic rafts, other floatation craft.

(*d*) *Rehearsals*. The patrol leader's order should include a plan for rehearsal. The plan should include a full-dress rehearsal on terrain similar to the objective, as well as a test firing of all weapons to be carried. An inspection should be conducted prior to departure to insure that all equipment, personnel, weapons, and attachments are functioning properly. The patrol leader should insure that patrol members do not have in their possession items of intelligence value, e.g., letters, marked maps.

(4) *Conduct of patrol.*

Passage of patrols through friendly positions. Coordination for passage should be made prior to patrol's arrival at the frontlines. Final coordination is made at the friendly command post and includes:

(*a*) Latest information of enemy.

(*b*) Final check on locations of wire, mines, boobytraps, etc.

(*c*) Size of patrol.

(*d*) Time of departure and return.

(*e*) Challenge and password and communications support.

(*f*) Relay of instructions about patrol to relieving personnel.

(*g*) Special or alternate recognition signals.

(*h*) Return of casualties or PW's through friendly positions. The patrol leader or assistant should count the patrol as it departs the friendly position to insure all personnel are present and in position.

(5) *Actions at objective.* Should be thoroughly rehearsed and understood by each member of the patrol. Alternate routes will be used only when the original route has been compromised.

(6) *Return to friendly lines.* Patrols nearing friendly outposts should move with extreme caution to avoid:

(*a*) Being fired on by outposts.

(*b*) Compromising challenge and password.

(*c*) Enemy patrols infiltrate friendly positions by joining patrol. Only the patrol leader and one other man should move to the outpost to establish identity. Each member of the patrol will be identified by the patrol leader as he passes the outpost. Patrols will report their return to the friendly unit through which they pass and proceed to debriefing headquarters as quickly as possible.

(7) *Wounded.* Render first aid and direct walking wounded to return by the designated route. Request assistance from higher headquarters for evacuation of wounded who are unable to walk. When behind enemy lines, move wounded to a concealed position of safety and comfort, mark location and attempt to bring them back on return trip. However, patrol mission has priority over disposition of wounded.

(8) *Prisoners.* Prisoners will be taken only if such action does not jeopardize the mission. When taken, they may be bound, gagged, and hidden for pickup on the return trip. If not picked up, their exact location will be reported on return.

(9) *After action.*

(*a*) *Debriefing.* The S2 will conduct a debriefing immediately upon return to the command post area. Patrol leaders and members will give an oral report. Detailed questioning can either add to or confirm known information.

(*b*) *Reporting.* A complete report may be prepared by the patrol leader for higher headquarters. This is normally done by the S2. Format designated by figure 44.

18. Physical Conditioning

a. Objective. To develop physical fitness, promote esprit, and to increase discipline and morale.

b. References. TM 21–200 and FM 21–20.

c. Requirements.

(1) *Cadre.*

(*a*) One principal instructor.

(*b*) One assistant instructor per game.

(*c*) One assistant instructor per running group.

(2) *Vehicles.* One ambulance.

(3) *Equipment.* Appropriate to game selected.

(4) *Training aids.* Appropriate to games selected.

(5) *Communication.* As required for games.

d. Instructor's Notes.

(1) *Introduction.* Organized physical training consists primarily of cross-country runs. Conditioning exercises are given prior to the run to warm the muscles, loosen joints, and stimulate blood flow. Mass games are conducted periodically after conditioning exercises to stimulate morale and encourage competition.

PATROL REPORT

A. Size and composition of patrol.

B. Task.

C. Time of departure.

D. Time of return.

E. Routes (out and back).

F. Terrain.

G. Enemy.

H. Any map corrections.

I. Miscellaneous information.

J. Results of encounters with enemy.

K. Condition of patrol.

L. Conclusions and recommendations.

Figure 44. Patrol report.

(2) *Physical conditioning.*

(*a*) *Toughening stage.* Use a moderate beginning to allow body to adjust to physical requirements. A minimum number of repetitions of drill one (TM 21–200) is a moderate beginning for warmup exercises. Runs should start with one-half mile distance or more.

(*b*) *Slow improvement stage.* The number of repetitions should increase gradually and the runs lengthened to four miles.

(*c*) *Sustaining stage.* When the peak is reached, conditioning is sustained through use of vigorous activities such as speed runs, mass games, and log exercises.

(3) *Exercises, runs, and games.*

(*a*) Exercises must be conducted vigorously and with precision to achieve coordination and timing.

(*b*) Exercises and runs must be supervised to correct errors and encourage Soldiers to complete the requirement. The Soldier who fails to complete the runs should be dealt with quickly, emphatically, and fairly. The failing individual should be encouraged to continue unless he is ill, injured, or unconscious. If he is unable to continue, additional exercises should be prescribed to increase his endurance. Persons unable to keep pace with other members of their groups are separated from the unit as soon as possible.

(*c*) There are many vigorous mass games, combatives, and relays (TM 21–200) which will add variety, enthusiasm, and interest to physical training programs.

(*d*) Games designed for maximum body contact may be made up to add originality as well as variety (1 fig. 45). Construct a circular hole 1.8 meters (six feet) deep and 2.7 to 3.6 meters nine to twelve (feet) in diameter. Two teams (any number of men) enter the pit and on a starting signal attempt to throw out members of the opposing team. A man is considered "out" of the pit when his feet are above or touch the top rim of the pit. The team which ejects the most members of the opposite team prior to the game's end is the winner. Water or mud in the pit will not dampen the enthusiasm of participants. The opposite of the Bear Pit is the Bunker Hill. In this game a mound of sand or mud similar in dimensions to the pit is used. The object is to pull, push, or drag the opposing team off the hill.

19. River Crossing Expedients

a. Objective. To acquaint the patrol member with some river or stream crossing expedients.

b. Requirements.

(1) *Troop.*

(*a*) *Cadre.* One principal instructor, two assistant instructors.

(*b*) *Support.* One boat man, two safety swimmers, and one aid man.

(2) *Vehicles.* One ambuance.

(3) *Equipment.*

(*a*) Machete.

1 The bear pit

2 Log exercises promote teamwork and coordination while physically developing personnel

Figure 45. Games and exercises.

(*b*) Other equipment brought to class by Soldiers, i.e., ponchos, rifles, survival ropes, etc. (needed for construction of raft).

(*c*) Safety boat.

(*d*) Safety rope.

(*e*) One two-man pneumatic reconnaissance boat.

(*f*) 30.48 meters nylon rope, *7/16*-inch.

c. Instructor's Notes.

(1) *Introduction.* This phase of instruction teaches the individual to cross wide, shallow streams, deep streams, and swamp

rivers. All Soldiers should be qualified swimmers (minimum of 50 meters).

(2) *Wide, shallow streams (less than 1.5 meters deep).*

(*a*) Select secluded, unlikely, shallow spots to ford.

(*b*) Post security and reconnaissance opposite bank.

(*c*) If depth is over ½ meter but under 1.5 meters, move patrol rapidly across in column, using a safety rope secured on both banks. In winter or during cold weather, the patrol should remove all clothing, cross hastily and dress quickly on the opposite shore. Security should be kept until every man is across. Weapons will be kept dry. This technique (one-rope bridge) may also be used when water is over 1.5 meters deep.

(3) *Streams and rivers (more than 1.5 meters deep).*

(*a*) *Two-man pneumatic raft.*

1. Raft may be used to shuttle equipment by securing the raft fore and aft with ropes tied to each shore.

2. Raft may be used to shuttle entire partol. Time required to shuttle all patrol members presents a disadvantage.

3. Raft may be used to carry weapons; partol members hold sides of the raft and are towed across from the far shore. This is a quick method of crossing. Rear security are last personnel to cross stream.

4. In cold weather, patrols should keep clothing and equipment dry. Personnel either ride raft across or carry clothing across on a raft.

(*b*) *Poncho raft.* This raft is used when no other flotation methods are available.

1. Seal poncho necks with draw strings.

2. Lay out ponchos one on top of the other, hood to hood.

3. Snap ponchos to each other along sides.

4. Cross rifles with fixed bayonets in scabbards diagonally on ponchos. (Cut poles may also be used when contact with enemy is imminent.)

5. Secure rifles at upper sling swivel with survival rope.

6. Stretch poncho taut diagonally and fold corners under rifle, leaving approximately 15.24 centimeters (6 inches) exposed.

7. While poncho is held taut at front, fold exposed poncho over the rifle butt three ways: first forward, then from the rear, then from the side.

8. Secure the folded poncho to the rifle with a bootlace.

9. Pad the bayonet affixed to the rifle with a pair of socks, fold the poncho as prescribed above and secure it with a bootlace.

10. Repeat the above operations to secure the remaining corners of the poncho. One bootlace is used to secure the front corner of the raft.

11. Wedge two combat packs containing a pair of boots each, between the poncho and crossed rifles to create the raft's hull. Thread the survival rope through the loops in each item.

12. Thread rope through the grommets at the front and rear, pull taut, and secure.

13. Thread the survival rope through the poncho grommets at the side, pull taut, and secure.

14. Join snaps that meet at the top of the raft and tuck away excess poncho.

15. Secure an empty canteen to the survival rope and tuck away excess rope.

16. Tie a bootlace with thumb-size bowline to the upstream front corner of the raft as a towline.

(4) *Soldiers practice construction and use of poncho raft.*

20. Survival

a. Objective. Acquaint Soldiers with methods of survival in all types of terrain and under all weather conditions. Instruction includes identification of edible plant and animal life, construction of shelters, and survival techniques.

b. References. FM 21–76.

c. Requirements. Requirements will vary depending on subject matter selected, locality, season, and availability of plant and animal life.

Figure 46. Poncho rafts afford the squad a limited amphibious capability. Nonswimmers and weak swimmers may use this technique under close supervision.

Figure 47. Overcoming the Soldier's natural fear of snakes also acquaints him with another survival food source.

d. Discussion.

(1) These classes are designed to provide information that could be a necessity on tomorrow's battlefield. Resupply of isolated strong points may prove to be just as impossible as the aerial resupply of General Wingate's operations in Burma during World War II. Here, aerial resupply proved inadequate despite the prior planning that preceded this operation.

(2) The principal instructor may lengthen the time of the final problem phase and require his Soldiers to employ the techniques taught during these periods of instruction. For best results, a Ranger training program should employ the use of survival type canned rations where the Soldier supplements his diet off the land. This technique assures an adequate diet, yet permits ample use of survival techniques.

e. Instructor's Notes. Appendix VI.

APPENDIX IV

CONDUCT OF PROBLEMS REFERENCED IN SCHEDULING

1. Day and Night Reconnaissance Patrol (Problem #1)

a. Objective. To acquaint the Soldier with the preparation, planning, and execution of a day and night reconnaissance patrol.

b. Reference. FM 7–15.

c. Requirements.

(1) *Cadre.* One principal instructor, one assistant instructor, one lane instructor per lane.

(2) *Support.* One aid man, one aggressor commander, one friendly frontline commander per simulated unit, personnel to support friendly and aggressor play.

(3) *Vehicle.* One ambulance, other vehicles as required to transport instructors, aggressors, and Soldiers.

(4) *Equipment.*

(*a*) *Soldier patrol.* See paragraph 26.

(*b*) *Aggressor.*

1. Demolition equipment.

2. Aggressor designation signs.

3. Automatic weapons w/blank adapters per lane.

4. Pioneer equipment, barbed wire, pickets, etc.

5. Snakebite kit.

(*c*) *Friendly positions.*

1. Automatic weapons placements.

2. Maps.

3. Pioneer equipment, barbed wire, pickets, etc.

(5) *Training aids.*

(*a*) Chart blow-up of problem area.

(*b*) Overlay showing disposition of enemy and friendly troops.

(*c*) Podium, pointer, blackboard, chalk, and eraser.

(*d*) Maps, overlays, equipment lists.

(*e*) Briefing tent.

(6) *Communications.*

Unit	*No. and type radi*
(*a*) PI	One AN/VRQ–3
(*b*) API	One AN/VRQ–3
(*c*) Agg position	One per position AN/PRC–10
(*d*) Friendly frontlines	One per position AN/PRC–10
(*e*) Base camp	One AN/PRC–10

d. Outline.

Note. The remainder of this paragraph is written in detail form. It is pertinent to the remaining problems and should be used as a guide in your organization of these problems.

(1) The training group is broken down into 5- to 9-man groups and receives a centralized reconnaissance patrol briefing. Each Soldier then prepares a warning order. One individual in each group is selected to give his warning order and assumes the duties of the patrol leader. In addition to complying with the instructions outlined in the warning order, each Soldier prepares a patrol order. The lane instructor may, if he desires, develop a situation that requires a change in the patrol leader any time during the play of the problem. The Soldiers are briefed to expect these situations and are encouraged to be prepared to assume this duty at any time.

(2) The problem is developed throughout the first two days of instruction by injecting combat reports received from a higher headquarters. These are summed into a final situation and given at the patrol briefing. An objective suitable for a reconnaissance patrol is selected and prepared prior to the Ranger training period.

(3) It may be necessary to modify the usual observation procedure if there is a limitation in qualified lane instructors. One lane instructor, through proper scheduling, may observe two patrols. This can be accomplished by staggering warning order and patrol order times. The friendly frontlines (FFL) commanders can prepare a critique on the Soldier's coordination and reconnaissance; the aggressor leaders in each lane can observe and list errors noted during the passage of enemy lines and at the objective. The friendly OP NCO can note discrepancies in coordination, and finally the lane instructor can critique both patrols by pointing out the errors made on the one patrol he observed. Inexperienced individuals generally make the same errors on their first patrol. Use this double patrol observation technique only as a last resort. See figure 48.

e. Phases.

(1) *Planning phase.* Stress patrol leaders action upon receipt of the order, his issuance of warning order, and selection of equipment to accomplish the mission. The planning phase is most important since it determines the conduct of preparation and rehearsal as well as the execution.

(2) *Preparation phase.* Evaluate the leader's organization of necessary preparatory measures, reconnaissance, issuance of patrol order, supervision of patrol preparation, inspection of

patrol and rehearsal. Evaluate the completeness and soundness of his plan. Evaluate subordinate leaders and members of the patrol in their performance of assigned precontribution to a team effort. The six **P**'s are most applicable: "**P**rior **P**lanning and **P**reparation **P**revent **P**oor **P**erformance."

Figure 48. "When we're on the march, we march single file, far enough apart so one shot can't go through two men." MAJOR ROBERT ROGERS–1756

(3) *Execution phase.* Evaluate the leader's ability to execute his plan and to control and command the patrol in the execution of the mission. Evaluate the attitude, cooperation, and actions of subordinate leaders and patrol members during the execution of the problem.

(4) *Critique phase.* Review the principles and techniques of patrolling involved in the exercise. Bring out the correct techniques used on the patrol, stressing development of strong points, and discussing the errors committed, stressing elimination of faults and weak points.

f. Exercise. A typical problem to familiarize personnel with the techniques and principles involved in the execution of any Ranger type field exercise follows:

(1) *General and special situations.* Background information relating to the tactical situation is presented by the principal instructor as a general and special situation. This information constitutes an operation order to the Soldiers and initiates the requirement for their initial actions. The situation may be presented in operations order sequence or as a series of messages and notes designed to require the student to assemble the information in operations order sequence. A suggested method of presenting the situation is—

(*a*) *General situation.* Suspected enemy positions are shown on the overlay. Exact positions are unknown. Morale unknown. Unit believed to be ____________ Division. Enemy has tanks, mortars, and artillery support. Air reconnaissance reports considerable activity in these areas. Our dispositions are as shown. Friendly plans are to continue the offensive, etc.

(*b*) *Special situation.*

1. Your company has been directed to send out reconnaissance patrols to determine the enemy strength, activity, and disposition in the battle group sector.

2. Your mission is to reconnoiter the high ground in your sector. The coordinates will be given later. You will study the terrain on your way out and on your return.

3. You will be taken to the forward company through which you will depart in time to make visual reconnaissance of the area and terrain over which you will move.

4. You will cross the line of departure (LD) at ____________ hours.

5. The latest information will be given to you at the company command post (CP).

6. The ____________ Field Artillery Battalion will be firing only on your objective until ____________ hours today. Firing will resume at ____________ hours.

7. Return through your friendly outpost by hours.

8. The coordinates of your objectives are ____________.

9. Casualties will be evacuated through the friendly outpost from which you depart. The patrol leader will determine final disposition en route.

10. One "C" ration meal will be carried by each man.

11. Higher headquarters furnishes transportation. Company furnishes ammunition and supplies.

12. I will receive your patrol report here when you return.

13. No radios will be carried. Challenge and password is ------------.

14. The time is now . . . are there any questions?

(2) *The patrol.*

(*a*) *Conduct of the patrol.*

1. The patrol is oriented in a location designated as a reserve battle group area. This orientation should be conducted by an individual designated as an S2. It includes the purpose of the problem, necessary safety and administrative details and a presentation of the general tactical situation. Following orientation of the patrol, the patrol leader may receive an operations order to include his patrol's specific mission. Maps, overlays, aerial photographs, and terrain models may be used in conjunction with the issuance of the operation order; however, the types and quantities of the visual aids employed are consistent with those normally found under similar circumstances in a combat situation.

2. The patrol leader is now responsible for the execution of the proper steps in troop leading procedure required for the successful accomplishment of his mission. Generally, his actions follow the sequence contained in the word picture of the problem below.

(a) The patrol leader makes an initial estimate of the situation and decides on an efficient way to utilize the time available to him. He makes a thorough map study and formulates a preliminary plan. He issues a warning order so that preparations can be started by all members. He makes any necessary coordination with personnel available in the area. Upon completion of this coordination he departs for the forward company through which his patrol will pass. He coordinates with the company commander through which his patrol will pass and obtains any information that may be available. He is then guided to a forward outpost where he makes visual reconnaissance of the terrain between his lines and the objective. He discusses pertinent details with the individuals in the outpost. He checks the progress of the patrol's preparation and makes changes in his preliminary plan, if necessary. He then completes his plan and prepares his patrol order. At the time and place designated in his warning order, he meets with the patrol and issues his patrol order. After answering questions and satisfying himself that his men understand the mission and their respective duties, he dismisses them to complete their preparations. He supervises the patrol and conducts any rehearsals he feels are necessary, such as formations that may be used and review of signals to be used. When possible, rehearsals are conducted on terrain similar to that over which the patrol is to operate. Prior to departure time, the leader conducts a final inspection. At the designated time, the patrol is moved forward to the forward company command post area. There, the patrol leader makes a final check with the company's commanding officer. Then, he and his patrol are guided forward to the outpost. The patrol leader checks with the personnel in the outpost for any last minute information they may have. He moves his patrol out and guides them on to his selected route. He utilizes pace men, point men, and compass men as desired. He adjusts or changes his formation based on the terrain, cover and concealment, visibility, and proximity to known or suspected enemy positions.

(b) When the patrol nears the enemy lines, the patrol leader selects a suitable area for the security teams. He leaves the previously designated security group in position and moves out on his reconnaissance with the individual or individuals he has designated to accompany him. The special situation imposes a time restriction on the patrol leader. He may move freely in his assigned objective area up to a certain time. At this designated time, he must be clear of the area due to friendly artillery shifting back to the objective. Reconnaissance completed, the patrol leader and his group rejoin the security element. At this time all information obtained is passed on to all members of the patrol. The patrol now moves out on its return route. Approaching the designated point of return in the friendly lines, the patrol leader slows down the patrol and moves cautiously until contact is established and recognition accomplished. The patrol passes through the friendly unit and returns to the company area. The patrol leader makes an immediate check of all personnel to see if anyone has information that should be included in his report.

(c) To provide training for all personnel, the patrol leader may be debriefed in the presence of the entire patrol. The patrol leader makes his report and the lane instructor subsequently conducts a critique. Following the critique, the LI counsels each individual as he deems necessary or appropriate. This may be accomplished as part of the patrol critique.

(*b*) *Post critique.* Upon completion of the counselling by the lane instructor, the Soldier should immediately readjust to the tactical situation at hand and be prepared mentally to assume the responsibilities of any forthcoming patrols. The Soldier should not be allowed to lapse into an attitude of administrative breaks between patrols. The problem play should be continued with periodic briefings and combat preparations. After a minimum period of time, during which the Soldier is required to maintain his clothing, equipment, and weapons as well as rest, an inspection is conducted to determine his fitness for another patrol. Immediately following the inspection, another orientation is conducted, and a new patrol mission assigned. Types of patrols should vary in scope and mission. Examples are included in the following problems.

2. Night Combat Patrol (Problem #2)

a. Objective. To develop the small unit leader's ability to organize and conduct a combat patrol; familiarize the Soldier with the principles and techniques of night combat patrolling; develop the individual's ability to infiltrate enemy battle positions, traverse his rear areas unobserved, destroy installations, disrupt communications, and return safely to the friendly lines.

b. Reference. FM 7–15.

c. Requirements.

(1) *Cadre.* One principal instructor, one lane instructor per lane, one assistant instructor, one aggressor commander, one friendly frontline commander, necessary personnel to support friendly frontline and aggressor play to include radio operators, fire direction center (FDC) personnel, and demolitions men, aid men, etc.

(2) *Vehicle.* ¼-ton for PI, API, FFL, (one per position) and one per aggressor commander. 2½-ton as required to transport aggressors and Ranger Soldiers.

(3) *Equipment.* For Soldiers, see checklist, paragraph 26.

(*a*) *Aggressor.* Demolition battery, wire, and other demolition equipment; insignia; snakebite kit, miscellaneous pioneer equipment; light machinegun per lane.

(*b*) *Base camp.* Snakebite kit.

(*c*) *Frontline position.* Light machinegun, maps, snakebite kit, demolition equipment, two candles.

(4) *Training aids.* Blowup of problem area, overlay showing disposition of friendly and aggressor troops, podium, pointer, equipment list, blackboards, chalk, eraser, maps, briefing tent.

(5) *Communications.*

(*a*) AN/VRQ–3 radio. One per principal instructor and assistant instructor.

(*b*) AN/PRC–6 radio. Three per patrol.

(*c*) AN/PRC–10 radio. One per aggressor position, two per FSCC, one per friendly frontline position, one per base camp, one or two per patrol.

d. Outline.

(1) The training groups of 18–22 men receive a centralized briefing. Each Soldier prepares a warning order. One individual in each group is selected to give his warning order and assume the duties of the patrol leader. In addition to following the instructions outlined in the warning order, each Soldier prepares a patrol order. The lane instructor develops situations that require a change in patrol leaders during the course of the patrol.

(2) This problem is based on the previously developed combat situations and should be coordinated with the mission results of the reconnaissance of previous patrols. An objective suitable for a combat raid is selected and prepared prior to the Ranger training period.

(3) Obstacles, both from the terrain encountered and enemy action, should be designed to periodically harass the patrol and its leader. Neither should be difficult enough nor severe enough to deny success to the operation.

e. Instructor's Notes.

(1) A simple plan, thoroughly rehearsed and vigorously executed, utilizing the element of surprise, offers the best chance for a successful raid.

(2) Night navigation is dependent upon a thorough map reconnaissance, the use of checkpoints, the recognition of prominent terrain features, and most important use of the compass.

(3) Reconnaissance, stealth, and covered movement by bounds are basically necessary when penetrating an enemy FEBA.

(4) Prior to assaulting the objective, a reconnaissance is conducted so that the original plan may be altered IF NECESSARY. It should be noted that altering plans endangers successful performance of the raid, and should only be done if success is denied, or surprise is lost, prior to arrival in the objective area.

(5) The raid plan should be flexible enough to permit rapid execution of preplanned alternatives. For instance, if a

sentry is to be eliminated in order to gain a swift silent accomplishment of the raid and the sentry is permitted to shout an alarm, the patrol must adopt the technique of firepower and shock to accomplish the raid. Such alterations should be planned.

(6) The patrol leader should present his information to the debriefing officer in as clear, concise, and complete a manner as possible.

3. Ambush Patrol (Problem #3)

a. Objective. To train troops in the proper techniques of night patrolling, conduct of clandestine assembly area, conduct of enemy convoy ambush and roadblock techniques.

b. Reference. FM 7–15.

c. Requirements. The requirements for this problem are very similar to problem #2, Night Combat Patrol. Variations, due to the mission, will be noted in the equipment lists used by patrol leaders. Vehicles used in the ambushed convoy may be the same used to transport Soldier and/or aggressor personnel.

d. Outline. Appendix IX, Example Problem Outline to a Training Memorandum. Appendix XII, Example Patrol Order.

e. Development. Patrol problems preceding the ambush patrol contribute to the changing friendly and enemy situation. Prior to the ambush patrol, a logical problem to be conducted is the prisoner raid. Information received from such a raid should then direct the play of the problem to the conduct of an ambush. If the Soldier is given access to the information which suggests a lucrative ambush target, he will normally think ahead, expect an ambush patrol and quite naturally retain a greater interest in the play of the problems.

f. Critique. Appendix XI.

4. Long-Range Raid To Seize and Hold Enemy Installation (Problem #4)

a. Objective. This patrol is designed to acquaint the Soldier with medium range patrolling, establishing a clandestine assembly area, use of rendezvous points, objective reconnaissance, methods of dealing with friendly agents, aerial resupply and link-up operations.

b. References. FM 7–15; FM 21–50, chapter 4.

c. Discussion.

(1) The problem requires infiltrating through the enemy's battle position in small groups, assembling at a rendezvous point behind enemy lines, moving to a clandestine assembly area, reconnoitering the objective, and seizing and defending the

objective until evacuated by air or until a link-up with friendly forces is effected.

(2) The problem is flexible and should incorporate all support that is available to the unit undergoing training. Airborne units can drop behind enemy lines in the play of the problem. A division reconnaissance company can be used as the link-up force. The objective can be a key piece of terrain or an important enemy atomic weapons fire direction center. It is visualized that the breakthrough of friendly forces will be preceded by an atomic strike.

(3) It is recommended that the distance for movement in this exercise be approximately 13 kilometers; however, the distance may be scaled to suit the available terrain.

(4) This problem is scheduled when the Soldiers are extremely fatigued and will answer the question: "How will my men perform under the most demanding conditions, when fatigued in combat?" If this problem is successfully organized, the members of the patrol (platoon size or larger) will receive sufficient rest in the clandestine assembly area.

d. Requirements. The requirements for this problem will vary too greatly to be listed in detail. The principal instructor or planning officer must determine the scope of the problem, to include terrain covered, duration, etc. The requirements and checklists used in other problems may be used as a guide in preparing a list of requirements.

5. Night Raid on Enemy Lines To Obtain Prisoners (Problem #5)

a. Objective. To develop the ability of leaders to prepare and execute under realistic combat conditions a thoroughly planned patrol. To develop the ability to seal off a portion of the enemy's battle position by use of supporting fire and to move into the sealed off area, attack his position, capture prisoners and withdraw.

b. Reference. FM 7–15.

c. Requirements. Same as problem #2. Variations may be made according to condition of terrain, weather, etc.

d. Outline. This problem is similar to the night combat patrol. The principal variation is that a "prisoner team" will physically capture prisoners. In the planning and rehearsing phases of the patrol, emphasis must be placed on prisoner handling; movement with prisoners inside enemy territory and when approaching friendly lines; selection and rehearsal of the prisoner team. This problem which is based on the previously developed combat situation demonstrates the importance of gaining information from prisoners. The information obtained will pertain to convoy movement by the enemy and will develop the situation so that an ambush patrol mission will logically follow.

6. Combat Raid Against Guerrilla Forces (Problem #6)

a. Objective. To familiarize Soldiers with techniques involved in a complex raid at night on an enemy guerrilla camp. Problem includes stream and cliff obstacles and movement by aircraft.

b. References. FM's 7–15; 31–21; chapter 4; appendixes XII, XIII, XV, XVI.

c. Requirements.

(1) *Cadre.* One principal instructor, one assistant instructor, one lane instructor per lane, ten enlisted assistant instructors.

(2) *Support.* Two aid men, one aggressor commander, five aggressors for each objective used.

(3) *Vehicles.* Two ambulances, other vehicles as required to transport instructors, aggressors and Ranger-Soldiers. Sufficient helicopters or fixed-wing aircraft to evacuate patrols.

(4) *Training aids.* Sufficient rope to breach stream and cliff.

(5) *Communications.*

Unit	*No. and type radios*
(*a*) PI	One AN/VRQ–3
(*b*) API	One AN/VRQ–3
(*c*) Control CP	One AN/VRQ–3 One AN/PRC–10
(*d*) Agg psn	One AN/PRC–10 per position
(*e*) Friendly frontlines	One AN/PRC–10 per position

Note. These radios are moved with cadre to major stream crossing sites as required.

(*f*) Base camp	One AN/PRC–10
(*g*) Patrols	One AN/PRC–10 per patrol

d. Outline. Sequence of Events: (See app. XIII).

(1) Soldiers are broken into patrol groups of approximately 25 men each.

(2) Soldiers move to patrol planning areas where warning orders are prepared based on S3 briefing. A patrol leader is designated and the warning order issued.

(3) Patrol moves through normal preparatory phase as in other problems to include drawing of equipment, coordinations, issuance of patrol order, rehearsals, and final inspection. During rehearsal helicopters should be made available with aircraft personnel to brief patrol on aircraft and loading; pathfinders should brief patrol on aircraft and loading; pathfinders should brief patrol on their procedures.

(4) After the planning phase, Soldiers are moved to friendly frontlines by truck and then on foot to the OP.

(5) The patrols move cross-country to the major stream crossing site where contact is made with a friendly partisan who provides a rope already implaced across the stream. After

crossing the stream a second rope is knotted and placed on a cliff to assist in movement up the cliff.

Notes. 1. If Soldiers are proficient in mountaineering expedient construction and operation, partisan should only supply rope; however, if Soldiers have received no mountaineering training, expedients should be constructed by partisan (cadre) prior to their arrival.

2. If cliff is not near stream, partisan could be used to guide patrol between sites.

(6) Soldiers then move to objective area and conduct combat type raid on guerrilla installation. Attack is basically a coordinated attack against a guerrilla force completely off guard. Objectives can include ammo dump, POL dumps, commo shacks, CP shacks, chemical dumps, and any other hasty type installations.

(7) Soldiers move immediately to helicopter pickup point for evacuation. Pathfinder team has secured landing site and is prepared to guide incoming aircraft.

(8) Soldiers return to friendly (rear) area and attend debriefing session. After eating and cleaning, and within 12 hours of return time, Soldiers attend general, patrol, and individual critiques.

e. Instructor's Notes.

(1) Basically, this patrol is a combat raid with the difference being that the enemy is a guerrilla organization. It has two major obstacles, a major stream and a cliff; helicopter evacuation adds complexity to the problem. Other actions, to include road and small stream crossings, enemy ambushes, etc., may be included or used on an "in lieu of" basis in this problem as desired. Distances may be increased or decreased, as desired.

(2) Soldier vehicles (2½-ton trucks) are on a standby basis at control CP for movement to nearest point outside guerrilla area which was predesignated for truck pickup, if helicopters could not land. Pathfinder should be in contact with control CP at all times.

(3) A critique checklist will insure each instructor covers the complete patrol period; at the same time, emphasis will be placed on the primary points desired.

7. Patrol Walk Through (Problem #9)

a. Objective. To acquaint the Soldier with the fundamentals required in the preparation, planning, and execution of a patrol.

b. Reference. FM 7–15.

c. Requirements.

(1) *Cadre.* One principal instructor and assistant, assistant instructor per lane.

(2) *Support.* One aid man, one aggressor commander, personnel to support friendly and aggressor play.

(3) *Vehicle.* One ambulance, other vehicles as required to transport instructor, aggressors, and Soldiers.

(4) *Equipment.*

(*a*) *Soldier* (may be simulated).

(*b*) *Aggressor.*

1. Automatic weapons w/blank adapters per lane.

2. Barbed wire, pickets.

3. Objective.

(*c*) *Friendly frontline.* OP emplacements, barbed wire, pickets.

(5) *Training aids.*

(*a*) Chart blowup of problem area.

(*b*) Overlay showing disposition of enemy and friendly troops.

(*c*) Podium, pointer, blackboard, chalk, and erasers.

(*d*) Patrol leader handout.

d. Outline. For best results this exercise is preceded by the last hour of Patrol Planning Orders and Techniques and is presented initially in the vicinity of the prepared patrol walk through lanes. The Soldiers are separated into 10- to 15-man groups and become part of a patrol organized and executed by an assistant instructor. The assistant instructor conducts his patrol as outlined in the following paragraphs after the students are acquainted with the patrol order.

e. Organization.

(1) Patrol member should receive written copy of a patrol order.

(2) Organize for actions at the objective in the following manner:

Patrol hq	Patrol ldr (lane instructor)
	Asst patrol leader
	Messenger
Assault element	Team leader
	Aid man
	Prisoner team
	Point
	Compass
	Pacers
Security element	Team leader
	Aid team
	Simulate two automatic weapons

Note. Designate elements to place out left flank security, right flank security, etc., as required. Give the order of march:

Point
Compass
Assault
Security

(3) Organization for march—

(*a*) *Emphasis.* Stress to the class that it is desirable to have a *security* element and an *assault* element.

(*b*) *Security element.* Heavy with automatic weapons if possible.

(*c*) *Assault element.* Normally controlled by the patrol leader. Does not have to include the entire assault element that will be used on the objective.

(*d*) *Objective.* Any members of patrol that have been moved from an element for the march to the objective can be moved back to their original element at the last halt before attacking the objective.

(4) Duties.

(*a*) *Point.*

1. Responsible for front security.

2. Function is not one of navigation.

3. Should be positioned far enough to the front to keep main body from being pinned down if hit.

4. Size of the point will vary with the size of the patrol and can be from one man to a squad.

5. The point maintains direction by observing the compass man.

(*b*) *Compass.*

1. Keeps patrol on the course designated by the patrol leader.

2. Generally located to the front of the patrol in the vicinity of the patrol leader.

3. Constant checks should be made by the patrol leader of compass man's adherence to route.

4. Additional compass men may be appointed to check or assist the primary compass man, but responsibility for the route still rests with the patrol leader.

5. The compass man should be located close to the point and within sight of the patrol leader and point.

6. Compass man has no security responsibility.

(*c*) *Pacers.*

1. Use two pacers and take the average from their totals.

2. Additional pacers may be used to pace from one march objective or checkpoint to another.

3. Pacers should be separated preferably one in the front and one to the rear of the patrol.

4. In small patrols when one pacer is used, the most accurate should be selected.

5. Patrol members with heavy bulky equipment should not be assigned as pacers.

6. The patrol leader should give definite instructions to adjust pace due to the terrain, etc.

7. Some methods of keeping pace are—

(a) Pace cord.

(b) Unbottoning flaps on cartridge belt.

(c) Putting pebbles in a pocket.

(*d*) *Assistant patrol leader.*

1. Generally located to the rear of the patrol, but is not the last man.

2. Primarily responsible for the rear security of the patrol.

3. Should send up the count after crossing each major obstacle and after all halts.

4. Responsibility to keep abreast of all plans, changes, etc. To do this it may be necessary to move up and down columns.

(*e*) *Miscellaneous.*

1. Messenger and radio operator should be located near the patrol leader for easy access.

2. Aid man should be a patrol member well qualified in first aid if no trained medical personnel are available.

3. Aid man should not have a key position which may require him to leave patrol's area.

4. Prisoner team is responsible for securing and handling of prisoners captured by the patrol.

5. Prisoner team is given the mission of securing or apprehending prisoners at the objective.

6. Demolition team is formed to handle special mission requirements which include use of demolitions.

Lane instructor notes. Review or emphasize the following paragraphs of the patrol order as pertains to your lane.

Mission.

Specific duties (most of which are covered in the organization above).

Routes to be followed.

Actions at danger areas (specifically what to do at suspected enemy position).

Actions at the objective.

Arm-and-hand signals.

f. *Rehearsal.*

(1) Cover the following points with group before you conduct a rehearsal:

(*a*) Reconnaissance of rehearsal area should be made prior to rehearsal to plan how you will utilize the area.

(*b*) Rehearsal should include all events as they will happen in sequence.

(*c*) Conduct of rehearsal. (This is one method.)

1. Walk through and explain to patrol layout of area, etc.

2. If time permits, allow each element and subelement time to rehearse.
3. Time permitting, have entire patrol conduct a walk through at half speed.
4. Finally, conduct at full speed.
5. Critique patrol after each phase of rehearsal.
6. Day and night rehearsals should be conducted if time permits.

(*d*) If time does not permit, priority of rehearsal should be:
1. Actions at the objective.
2. Actions upon enemy contact.
3. Danger areas.
4. Control and security.

(*e*) Assistant patrol leader can be used to set up rehearsal area by using #10 cans, engineer tape, etc., or to make a hasty sandtable.

Lane instructor notes. **Move the patrol to designated area to conduct a rehearsal of actions at the objective. (Explain to class because of time element, this is the only item being rehearsed so that they will know the difference.)**

g. Inspection.

Lane instructor notes. **Lane instructor will conduct an inspection using only a few members from each element to demonstrate what should be checked, etc. (see below).**

Try to use members from all of the elements to demonstrate the items the patrol leader would be concerned with in each group.

(1) Detailed inspection will include the following:

(*a*) Insure first leg of compass is set by compass man. Compass man has compass with first leg set on compass.
(*b*) Pacers—have pace cords (furnished by PI).
(*c*) Prisoner team—utility ropes for binding.
(*d*) Check all personnel for:
1. Colds.
2. Special items of equipment such as flashlight, first aid kit, grenade launchers, snakebite kits, etc.
3. Weapons for taped slings, stacking swivel.
4. Camouflage.
5. No shiny objects.
6. No items carried that will make noise.

(*e*) Outer garments or uniforms are as prescribed.
(*f*) All personnel should have pencil and paper.
(*g*) Cover method of carrying machinegun ammunition in pack.
(*h*) Stress the danger of carrying marked maps, personal letters, manuals, and notebooks with briefing.

(2) On large patrols, team leaders make detailed inspections before the patrol leader makes his detailed or spot inspection.

(3) If patrol leader departs from the area prior to patrol inspection, the assistant patrol leader will inspect the patrol in the reserve area. The patrol leader will later spot check the patrol before leaving friendly lines.

(4) During the inspection, the patrol leader asks questions of various patrol members about the patrol order to insure information has been absorbed.

LANE INSTRUCTOR MOVES THE PATROL FROM THE REHEARSAL AND INSPECTION AREA TO THE OUTPOST.

h. Coordination at the Friendly Outpost.

(1) Meet guide.

(*a*) Patrol leader gives identity to guide.

(*b*) Patrol leader checks distance to OP and has the guide inform him when the patrol is 50 to 100 meters from the OP.

(*c*) Patrol leader should question guide about the route for possible danger areas, etc.

Lane instructor notes. **Move entire patrol to CP-OP to observe coordination. (Explain to patrol that they are being allowed to move in close in order to hear the coordination.)**

(2) At CP-OP:

(*a*) Patrol leader identifies himself.

(*b*) Informs CO on size of patrol and mission.

(*c*) Informs CO time of patrol's return.

(*d*) Verifies password and challenge.

(*e*) Inquires about latest information on enemy small arms fire, location, time, etc.

(*f*) Inquires about friendly wire, boobytraps, etc., near position.

(*g*) Arranges for a guide to gap in friendly wire.

(*h*) Requests information on terrain to front.

(*i*) Checks to see how long personnel have been at OP, and if they will pass on information to relief if relieved before patrol returns.

(*j*) Asks permission to establish initial rally point near OP.

(*k*) Asks for support CO can give with small arms, mortars, alert squads, litter teams, etc.

(*l*) Navigational signals or aids company can provide.

(*m*) Information on friendly defensive positions.

(*n*) Inform CO of pyrotechnic plan (what color patrols will use, etc.).

(*o*) Coordinate rehearsal and mess area if needed.

(*p*) See if CO can enter patrol radio net and act as relay.

(*q*) If company has no spare radio, find out what frequency company is on.

(3) Stress to group that a list of questions should be made and kept in a small pocket notebook to use while coordinating phases.

(4) Discuss how the assistant patrol leader counts patrol through the gap and falls in on the rear of the patrol. The reasons for doing this—

i. Security Halt.

Lane instructor notes: **Patrol proceeds to the gap in the friendly wire. The lane instructor halts the patrol and directs a two-man reconnaissance at least 100 meters left and right and to the front on the enemy side of the wire. When the area has been screened, the two men meet and one will signal that all is clear. The entire patrol is moved through the gap until the end of the patrol is clear and then the lane instructor will have the patrol make a security halt.**

(1) *Explain to the group that we halt to*:

(*a*) Get accustomed to noises.

(*b*) Get acquainted with terrain and any unusual features of the area.

(2) *Stress*:

(*a*) This is the first time on patrol that the assistant patrol leader begins to function as he counts the patrol through and sends up count.

(*b*) A good technique is to have the entire patrol look back at the area so they will recognize it when they return.

(*c*) Everyone must be absolutely quiet with no shifting of equipment or movement when the patrol halts.

j. Movement and Conduct Between Enemy and Friendly Positions.

LANE INSTRUCTOR WILL CONDUCT THE PATROL IN A COLUMN WHERE POSSIBLE.

(1) Discuss the advantages of:

(*a*) *Column.* Firepower limited to front but good to flank; provides excellent control; good movement along roads, defiles and when visibility is poor.

(*b*) *Single file.* In the interest of speed and control through dense woods or swamps, the single file may be used. This formation is to be used to count patrol through wire or OP's. Flank security should be used as much as possible. Don't forget to change to a more suitable formation upon reaching an open area or in daylight.

(*c*) *Formations.* Any of the combat formations for the squad or platoon can be used.

LANE INSTRUCTOR HAS THE PATROL EXECUTE THE FOLLOWING DURING MOVEMENT AND CRITIQUES AFTER EACH IF NECESSARY.

(2) Passing up COUNT.

(3) Passing up PACE.

(4) Momentary HALT. Have patrol to either stand or drop to one knee in place and face directions of responsibility.

(5) Temporary HALT. Assistant patrol leader should place out necessary security and report for special instructions. Normally, the patrol will be in the prone position.

Note. **Cover NO SMOKING between main battle positions or in close proximity of enemy. Extreme caution should be used when smoking behind enemy lines especially if contact was made going through enemy lines.**

LANE INSTRUCTOR DESIGNATES SECOND RALLYING POINT. GATHER PATROL AND COVER.

(6) General information on rallying point:

(*a*) Select by map reconnaissance, confirm on the ground, then designate as patrol passes.

(*b*) Use well defined and easily located features as rallying points.

(*c*) Select rallying points as required by *distance*, *danger areas* and terrain.

(7) Actions at a rallying point.

(*a*) Maximum security is established immediately by senior in command.

(*b*) Wait the time specified in patrol order or by verbal instructions.

(*c*) Take a head count.

(*d*) Check and redistribute ammunition and equipment.

(*e*) Reestablish chain of command.

ACTION BELOW HAS BEEN EXPLAINED TO PATROL DURING ORGANIZATION AT INSTRUCTION SITE. (Actions at danger areas.) After this action the value of rehearsals will become apparent to patrol members.

(8) Patrol will be fired upon by an aggressor machinegun at a range of 25 to 75 meters.

(*a*) Immediately, all of the patrol returns fire. Security team leader moves weapons into good firing positions. Platoon leader (lane instructor) makes an estimate of the situation, followed by a plan of action.

(*b*) Plan is to withdraw assault element with the security covering followed by the withdrawal of the security. This will be continued until contact is broken.

(c) Patrol leader (lane instructor) throws grenade simulator in direction of aggressor fire and commands: "6 O'CLOCK 200."

(d) Assault element will break in the direction of clock described by lane instructor for actual distance of 50 meters (simulating 200 meters) and take up firing positions to cover the withdrawal of security.

(e) As soon as assault is in position, a signal is given; the security team led by assistant patrol leader will move to join the remainder of patrol.

(f) Patrol leader (lane instructor) regroups patrol and bypasses position.

Lane instructor note. Critiques action.

Note. It must be explained that ambush would require different actions and techniques.

(g) Discuss returning to last rallying point, if patrol becomes scattered and dispersed immediately upon receiving fire.

k. Crossing Roads and Streams.

(1) Roads—the patrol will cross a road south of the objective using reconnaissance and security. Reconnaissance elements will physically reconnoiter 50 meters to left and right and far side of road. The road will be crossed by squads or sections in line of skirmishers under cover of security.

Lane instructor notes. Discuss techniques of crossing at bends in road; crossing single file stepping in the footprints of the lead man; crossing in regular formation at night with adequate security.

(2) Streams—same principle as roads.

(3) Using roads—guiding on roads is normally not a good practice. If roads are used, point man should be well forward of patrol, connecting files placed out, and keep rear alert.

l. Actions at the Objective.

Lane instructor notes. As the patrol approaches the objective, the point is well forward of the main body. The point sights the objective and alerts patrol. The lane instructor places the patrol in a perimeter under command of the assistant patrol leader and moves forward on reconnaissance. Personnel accompanying the patrol leader in the reconnaissance party are the assault and security team leaders and the messenger. The lane instructor at this time, designates assault position, location of automatic weapons, etc. Upon return of the patrol, information gained is disseminated.

(LANE INSTRUCTOR SHOULD GIVE BRIEF REVIEW TO KEY LEADERS ON HIGHLIGHTS OF INFORMATION GIVEN IN THE PATROL ORDER ABOUT ACTIONS AT THE OBJECTIVE.)

Note. Be sure to designate this area or another area as an assembly area. The assault element moves at least 50 to 75 meters past objective before reorganizing. The assault team leader moves individuals into good firing positions, renders casualty, ammunition reports, etc. The lane instructor will tell the patrol he will destroy the objective with a grenade as the patrol withdraws. The prisoner team (with prisoner) will withdraw to the rallying point followed by the assault team with the lane instructor next. As the lane instructor passes the security element, he will tell the security team leader that the remainder of the patrol is en route to the rallying point. The patrol (–) takes up a perimeter upon arrival at the rallying point. As soon as the entire patrol is assembled, a head count is taken, ammunition redistributed, etc., and the patrol moves out on return route.

THE LANE INSTRUCTOR AT THIS POINT HALTS THE PROBLEM TO REVIEW AND CRITIQUE ACTIONS AT THE OBJECTIVE.

(1) He reviews actions of the patrol from the time the point sighted the objective *stressing his actions* as patrol leader.

(2) Review the functions of security and support.

(*a*) Sealing the objective area by fire.

(*b*) Covering the withdrawal of other patrol elements.

(3) Use of artillery if available. Artillery and mortar fire is used to soften objective, smoke or mark it, block routes of enemy escape or reinforcement. Support element should have the capability to shift supporting fires as the assault element reaches the objective.

(4) Prisoner team should have ropes, handcuffs, etc., ready for instant use. Securing a prisoner's neck, apply gag, and tie his hands in front so he can use them for climbing, etc.

(5) Movement out of the objective area should be orderly and controlled, not in haste. Ammunition should be redistributed by elements on the objective before leaving.

Lane instructor's notes. After questions and comments, cover the following (time permitting).

m. Return Route.

(1) Security is imperative since the enemy probably knows of your presence and is alert for the patrol's return.

(2) Approach friendly lines carefully; do not relax security.

(3) Do not parallel the friendly lines while searching for a point of entrance.

(4) Know the correct use of challenge and password. The patrol leader advances to OP, identifies himself, then counts his men through the outpost to insure all are present and no enemy infiltrated with patrol.

(5) Give a spot report on the CP-OP concerning terrain, enemy contact, etc.

n. Enemy Wire.

(1) Methods of crossing.

(*a*) *Bypassing.* This requires thorough reconnaissance and time. Prevents possibility of excessive noise while passing through wire.

(*b*) *Crawl under wire.* If wire construction permits, this technique allows leaving the wire intact and preventing enemy discovery.

(*c*) *Cutting wire.* This method allows quick passage through wire but may give patrol away to enemy. Always leave top strand of wire uncut to minimize chance of enemy detection.

(*d*) *Leaping over wire.* Quickest means of clearing wire. One may lay across wire which allows the patrol to use his body as a bridge to cross the wire. This method is noisy and should not be used in close proximity of enemy positions.

(2) Security should be placed on far side of enemy wire to cover passage of main body of patrol. Automatic weapons are moved forward to cover patrol crossing wire.

o. Suggested Time Schedule.

Bleacher inspection (Techniques of patrolling)	(50 min)	1315–1405
Patrol organization	(25 min)	1405–1430
Rehearsal	(20 min)	1430–1450
Inspection	(20 min)	1450–1510
Coordination at outpost	(20 min)	1510–1530
Security halt	(20 min)	1530–1550
Movement between lines	(25 min)	1550–1615
Actions at the objective and critique	(30 min)	1615–1645

p. Patrol Order.

(1) *Situation.*

(*a*) *Enemy forces.* Terrain is slash pine and scrub oak with thickets near the creek. Generally, there is good cover and concealment with a few open areas. All creeks in the area are fordable. Since this is a daylight mission, we will be concerned with the sun which will set at -------- hours. However, in the event we return late, there will be a moon tonight. We do not know the exact identity of the aggressor in the area, but they are believed to be from the 16th Fusilier Division. Known enemy appears to be manning all positions thinly, and is expected to withdraw to his new position west of OCHILLEE Creek at any moment. G2 estimates the aggressor is down to about 50 percent effective fighting strength.

(*b*) *Friendly forces.* There will be -------- patrols from our unit operating in the same general area. All patrols will move on parallel route. Necessary coordination has been made with all patrols. There is no fire support available for our mission.

(c) *Attachment and detachments.* None.

(2) *Mission.* Our mission is to attack the assigned objective and capture prisoners before the aggressor withdraws from his present positions.

(3) *Execution.*

(a) *General plan.* The patrol will proceed generally northwest to the vicinity of the objective where a patrol perimeter will be established, while a reconnaissance is made. We will assault our objective and return to the OP on the back azimuth with enemy captured on objective.

(b) *Specific duties of each individual or unit.* (To be covered by individual lane instructors.)

(c) *Coordinating instructions.*

1. Time of departure and return. Depart 1500 hours today from the OP of 1st Plat, Co A, 1/14th Inf; return NLG ---- hours this evening.

2. Route to be followed. (To be covered by individual lane instructor.)

3. Passage of friendly position. Patrol will move by foot from company area to the vicinity of 1st Plat, Co A, 1/14th Inf where it will pick up a guide who will lead the way to the CP-OP. We will go through the outpost in a single file. The guide will take us to the friendly wire; assistant patrol leader will count the patrol through after number 1 and 2 men have made a reconnaissance of the enemy side of the wire. We will have a security halt to become oriented after passing through the wire.

4. Alternate route of return. Utilizing azimuth from departure point to objective as a base, move on an azimuth of 90 degrees for 200 meters from the objective. Then, turn 90 degrees more (toward friendly lines) and move on this course. This azimuth will be 180 degrees from the initial azimuth from the FFL to the objective.

5. Actions of danger areas.

(a) At CP-OP, form a perimeter 100 to 200 meters in rear of CP-OP with 12 o'clock being the direction of movement.

(b) At creeks, roads, trails and open areas, number 1 and 2 men will cross first and check far side and flanks while the remainder of patrol forms a perimeter. If clear, signal me and we will cross in the manner prescribed by me at the time.

6. Initial formation. After passing through friendly wire, we will proceed in a platoon column with the assault leading, followed by patrol headquarters and security bringing up the rear. Watch my signal to change formations.

7. *Action upon enemy contact.* Enemy contact will be avoided if possible. If fired upon by the enemy, personnel in position to return effective fire will do so. The assault element will withdraw to a firing position to support the withdrawal of the security element. Fire and movement will continue until contact is broken, or we are out of effective range. We will use the clock system to break contact with 12 o'clock being the direction of movement. In the event of ambush, return fire immediately and attack in a skirmish line in the direction I indicate.
8. *Initial rallying point.* 75 to 100 meters to the rear of Co A, CP-OP. I will designate this as we pass.
9. *Actions at rallying points.* Senior in command set up security. After 25 percent of the patrol is present or 25 minutes after the first men arrive, continue with the mission.
10. *Action at objective.* (To be covered by individual lane instructor.)
11. *Reporting results of patrol.* Patrol will report to Bn S2 at 1st Ranger Bn Hq upon return.
12. *Rehearsals.* Rehearsals with all weapons, equipment and camouflage will be conducted at 1430 hours today.

(4) *Administration and logistics.*

(*a*) *Rations.* No rations will be carried.

(*b*) *Arms and ammunition.* (As prescribed by individual lane instructor.)

(*c*) *Special equipment.* (As prescribed by individual lane instructor.)

(*d*) *Method of handling wounded and prisoners.* En route to the objective, wounded will receive first aid and be picked up by patrol on return. After reaching the objective, wounded will be evacuated with the patrol. Prisoners taken en route to the objective will be bound, gagged, and hidden until return of patrol. Prisoners taken at the objective or on our return route will be handled by the prisoner team.

(5) *Command and signal.*

(*a*) *Signals.* We will use arm-and-hand signals whenever visibility permits. (At rehearsals I will show signals for halt, move into prone position, up on your feet, move forward, enemy sighted, commence firing, cease firing, etc.) Luminous compass during darkness; compass moved vertically means all clear, proceed; compass moved horizontally means halt, danger. Radio will be carried—

Prim 38.7	Call Sign : Blue 1
Altn 39.4	Bn is Red 6

Challenge and password: 1200 hours today until 1200 hours tomorrow—GREEN WOOD. 1200 hours tomorrow until 1200 hours next day—LONG BOY. Do not use challenge and password forward of our lines. The patrol challenge and password, or with other patrols from our company, will be HOT HANDS.

(*b*) *Chain of command.* (To be covered by individual lane instructor.) The time is now ------------ hours. ANY QUESTIONS?

APPENDIX V

EXAMPLE MOUNTAINEERING LESSON OUTLINES

Section I. ROPES AND KNOTS

TIME ALLOTTED:	1 hour.
TOOLS, EQUIPMENT, AND MATERIAL:	Nylon climbing rope, nylon sling rope, ¼-inch, ½-inch, ¾-inch and 1-inch manila ropes.
PERSONNEL:	1 principal instructor, 1 assistant instructor per 6 to 8 Soldiers.
SOLDIER UNIFORM AND EQUIPMENT:	Field uniform with sling rope, snaplinks and heavy leather gloves.

Note. Class area should be large enough to allow all Soldiers to tie knots on log corral or other suitable anchors. Area should provide place for Soldiers to practice throwing rope, preferably from heights.

PURPOSE: To familiarize Soldiers with use of ropes and knots in mountaineering.

Types and Characteristics of Ropes. The climber will find use for three different types of rope in his work.

1. Nylon climbing rope comes in coils of 120 feet in length and 7/16-inch in diameter. It has a breaking strength of approximately 3,600 pounds when new. It will stretch approximately 1/3 its length. Its great elasticity is valuable as a safety factor.
2. Manila sling rope is commonly used in 12- to 14-foot lengths. It is ¼-inch or more in diameter and should have a minimum breaking strength of 550 pounds. Nylon sling ropes are the same length as manila and have the same breaking strength as the nylon climbing rope.
3. Three-quarter-inch and one-inch rope are frequently used in the construction of various types of riggings requiring extensive spanning. This rope is better than nylon for suspension traverses and rope bridges because it has less elasticity. When suspended between two anchor points, 5 percent slack should be left in the rope.

Care of Rope.

1. Since the rope frequently is the climber's lifeline it deserves a great deal of care and respect.

2. The rope should not be stepped on or dragged on the ground. Small particles of dirt will be ground between the strands and slowly cut them.
3. The rope should not be in contact with sharp corners or edges of rock, as this will cut it.
4. Keep the rope as dry as possible. If it should become wet, dry it as soon as possible to prevent rotting.
5. Do not leave the rope knotted or tightly stretched longer than necessary, and do not hang it on sharp edges.
6. When using rope in installations, do not let one rope rub against another as this will cut and/or fray the ropes; however, many times this cannot be avoided.
7. Particular care must be taken with nylon rope as heat generated by rope friction will melt the fibers.
8. The rope should be inspected often for frayed or cut spots, mildew, and rot. If such spots are found, the rope should be whipped on both sides of the bad spots and then cut.
9. New sling rope, as well as any other rope that is cut from a long piece, should be whipped at the ends.
10. Climbing ropes should *never* be spliced.
11. Climbing ropes should be marked in the middle. This can be done easily by untwisting the rope at the middle and inserting a small piece of manila.

Coiling. Two methods of coiling rope are used.

1. An end of rope is taken in the left hand with the right hand running along the rope until both arms are outstretched. The hands are brought together forming a loop which is laid in the left hand. This is repeated, forming uniform loops, until the rope is completely coiled. If there is any tendency for the rope to twist or form figure eights, it may be given a slight twist with the right hand to overcome this. Right lay rope should always be coiled in a clockwise direction.
2. The rope is coiled around the foot and knee. In coiling on the left leg start from the inside, bring the rope over the knee to the outside and around the foot to the inside. Continue in this manner until the rope is coiled, using the same technique as in the previous coil.
3. Tying off coil. To tie the Mountain Coil, a bight a foot long is made in the starting end of rope and laid along the top of the coil. Uncoil the last loop; take the length of rope thus formed and wrap it around the coil and the bight. The first wrap is made at the open end of the bight in such a manner as to lock itself. Continue wrapping toward the closed end until just enough rope remains to insert through the bight. Pull the running end of the bight to secure the

wrapped rope. A rope properly coiled has from six (6) to eight (8) wraps in the tieoff.

Terms Used in Rope Work.

1. Bight: A bight of rope is a simple bend of rope in which the rope *does not* cross itself (fig. 49).

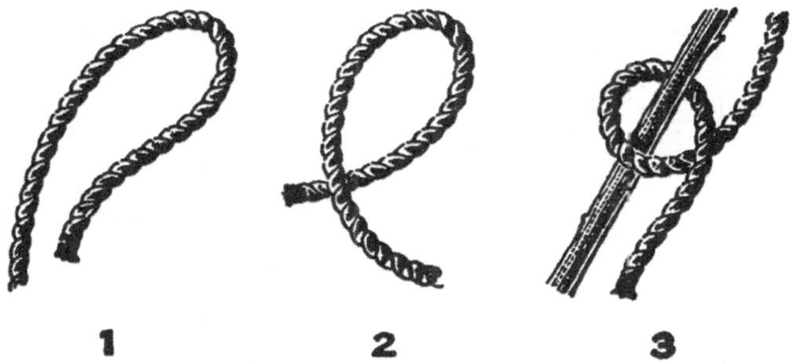

Figure 49. Bight, loop, half hitch.

2. Loop: A loop is a simple bend of rope in which the rope *does* cross itself (fig. 49).
3. Half hitch: A half hitch is a loop which runs around an object in such a manner as to lock itself (fig. 49).
4. The running end is the free end which can be used.
5. The standing part of the rope is the fastened part.
6. The lay of the rope is the same as the twist.

Knots.

1. *Square knot:* Uused to tie the ends of two ropes of equal diameter together. It should always be secured by a half hitch on each side of the knot. This knot will not slip and will draw tight under strain.
2. *Double sheet bend:* Used to tie the ends of two ropes together, whether of equal diameter or not. It can also be used to tie the ends of several ropes to the end of one rope (fig. 50).

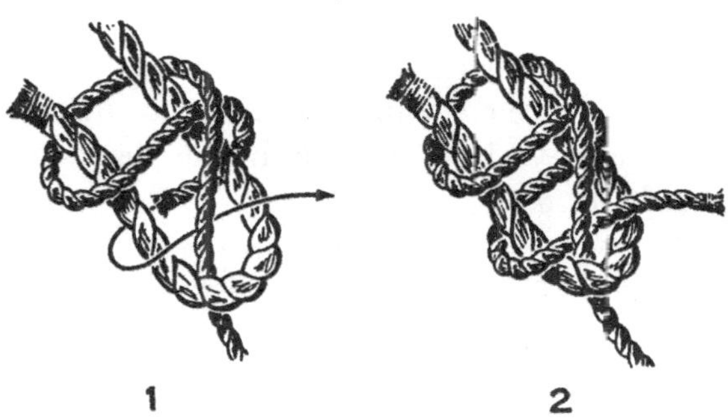

Figure 50. Double sheet bend.

3. *Butterfly:* Can be tied in the middle of rope without using the ends. May be used for middle man in three-man party climbs and also in tightening installation ropes (fig. 51).

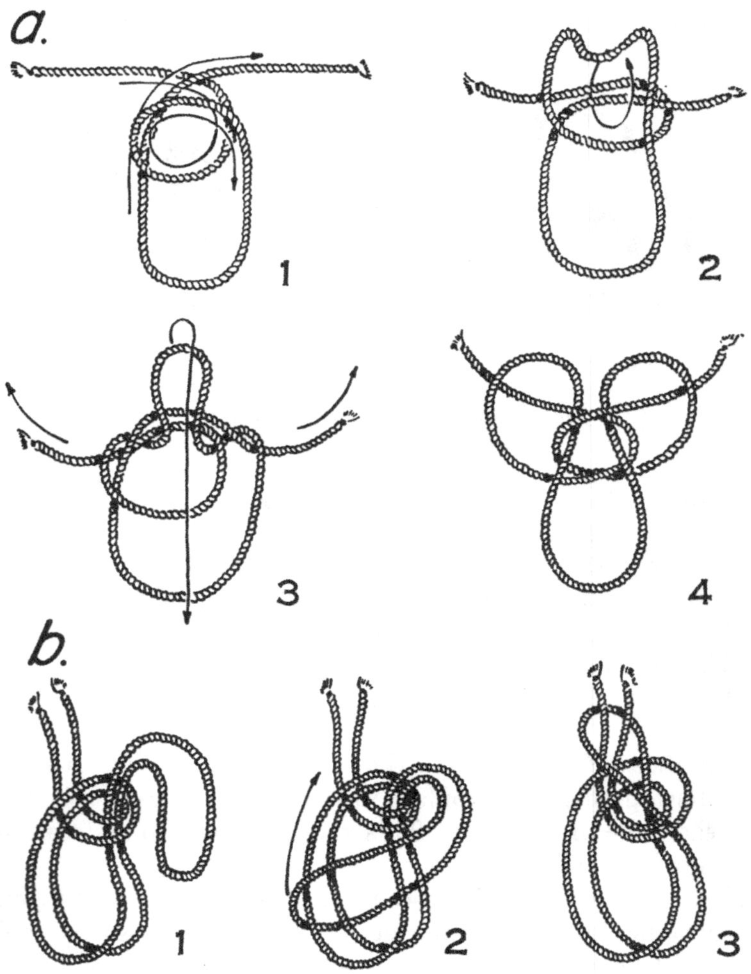

Figure 51. a. Butterfly knot. b. Bowline on a loop.

4. *Bowline:* Used for end man on a climbing rope and as an anchor knot where constant tension is not maintained. The bowline is always half hitched inside the main loop after knot is tied (fig. 52).
5. *Round turn with two half hitches:* A strong, easy to tie and untie anchor knot. Used only when constant tension is maintained (fig. 52).
6. *Clove hitch:* Used as an anchor knot when constant tension is maintained.

7. *Bowline on a coil:* Used by the end man on a climbing rope to take up extra and unnecessary slack (fig. 53).
8. *Prussik knot:* Used to anchor the fixed rope to various anchor points. It is also used to make crevasse rescues. It is tied with a bight of rope and also with an end of a rope. When tied with an end of rope, it is finished off with a bowline.
9. *Figure eight slip knot:* Used as an anchor knot on fixed rope.
10. *Transport knot:* Consists of a slip knot and a half hitch. It is used to tie off the tightening arrangement on rope installations.
11. *Bowline on a bight:* This bowline forms a double loop. It can be used for both tying in on climbing ropes and as a piton anchor using four (4) pitons.

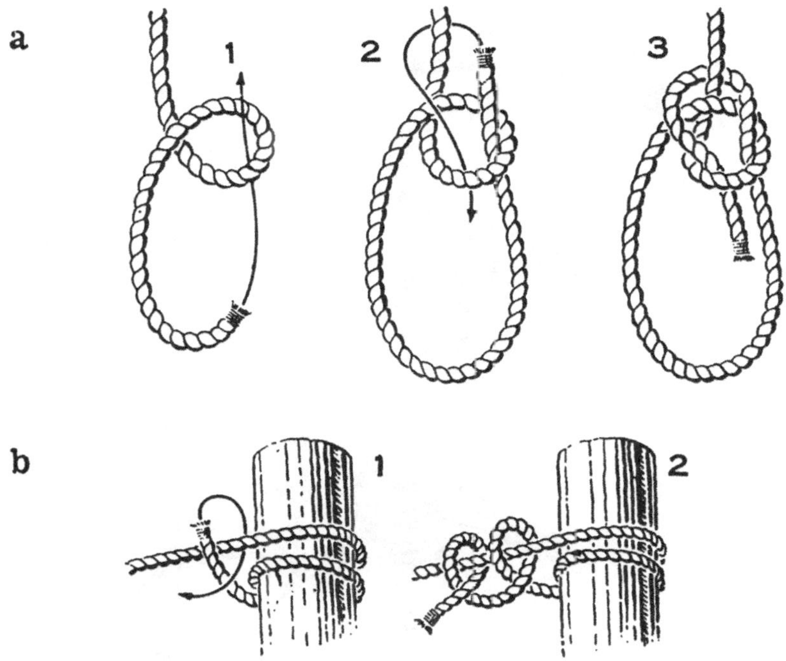

Figure 52. a. Bowline. b. Round turn with two half hitches.

12. *Three loop bowline:* This bowline forms three loops. It is used for both a piton anchor, using four (4) pitons, and as a seat in evacuation, etc. (fig. 53).
13. *Overhand knot:* This knot is used to make the carrying rope for a suspension traverse, stirrups for tension climbing and to make a knotted rope used as a handline to assist men on a vertical hauling line (fig. 54).

Rope Throwing. In most cases, it will be necessary to carefully recoil the rope before throwing. In throwing the full length of rope,

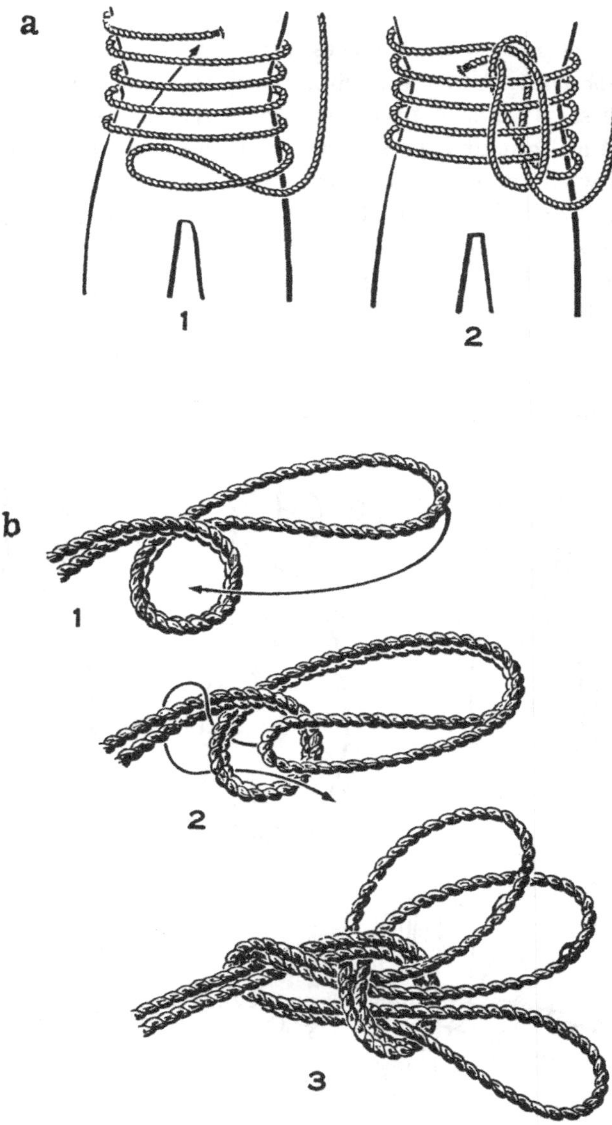

Figure 53. a. Bowline on a coil. b. Three loop bowline.

grasp the coil in the right hand, take the end of the rope nearest the fingertips and anchor it. Take five or six loops from this end of the coil and hold in the left hand retaining the remaining coil in the right

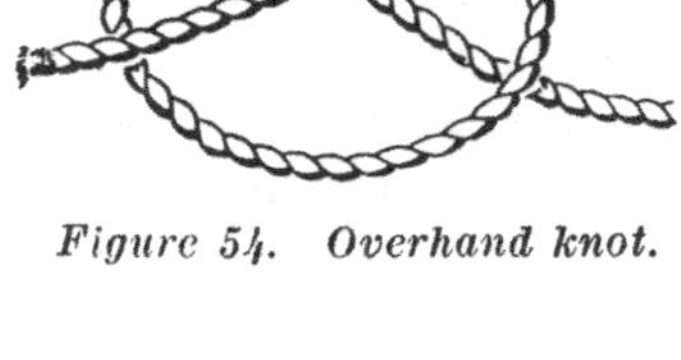

Figure 54. Overhand knot.

hand. The right hand coil is thrown first. A couple of preliminary swings will insure a smooth throw. The swings should be made with the arm partially extended and the coil should be thrown out and up. A slight twist of the wrist, so that the palm of the hand comes up as rope is thrown, will cause the coil to turn, the loops to spread, and the running end to fall free away from the thrower. A smooth follow through is essential. As soon as the coil is thrown and spreading, the few loops held in the left hand should be tossed out. The rope should be thrown into the wind so that the running end is to the leeward.

Section II. AIDS FOR ROPE INSTALLATIONS

TIME ALLOTTED:	Integrated in several instructional blocks.
TOOLS, EQUIPMENT, AND MATERIALS:	Deadman, picket holdfast, piton anchors, A-frame, climbing ropes and snaplinks.
PERSONNEL:	Principal instructor, 1 assistant instructor per 6 to 8 Soldiers.
INSTRUCTIONAL AIDS:	Fixed installation display area.
REFERENCE:	FM 31–72.
SOLDIER UNIFORM AND EQUIPMENT:	Field uniform w/sling rope, snaplink and heavy leather gloves.
TROOP REQUIREMENTS:	Medical aidman w/litter jeep.

Note. All points covered in instructional material attached are covered in one or more of the other instructional blocks pertaining to rope installations. Initial introduction to Basic Military Mountaineering Techniques includes description of various installations and pointing out of these installations already set up and functional.

One of the most important duties of the military mountaineer is to assist the unskilled men of his unit over difficult terrain. Aids for this work are fixed ropes, vertical holding lines, and suspension traverses or any other similar rope expedient. These aids may require the use of anchors, tightening knots, and A-frames.

Anchors. In the setting up of all rope installations, the problem of the main anchor is a great one. The ideal situation is to have some good natural object such as a firmly rooted tree or solid rock nubbin. Since this is not always available, anchors must be made or devised by artificial means. These are called "deadmen" and the leader of the installing party must decide which form is most efficient in regards to speed of installation, safety, and durability.

Natural anchors. Since these are always preferable, their use should be studied with care. If a tree is to be used, its firmness is of the greatest importance. This is especially true if the installation will be used for any length of time. Trees growing on generally rocky terrain should be treated with suspicion, as their roots normally are shallow and spread out along a relatively flat surface. If the tree has been found to be satisfactory, the rope may be tied to it with any one of the anchor knots. If the installation is to be used for a great length of time, the bowline or the round turn with two half hitches is preferable to the clove hitch. If rock nubbins can be used, their firmness is again of primary importance. They should be checked for cracks, or any other signs of weathering that may impair their firmness. If any of these signs exist on a nubbin, its use should be avoided, except in cases of absolute necessity, and then only after careful and thorough testing. Sharp edges will frequently be found on nubbins. These should be padded with extra clothing, rags, branches, or grass.

Artificial anchors. Artificial anchors can be divided into two main classes. There are anchors that are installed in earth or dirt and those that are put on rock with pitons. Artificial anchors in earth are of two types. The *single timber deadman* is the safest type, although it takes quite some work to construct. A trench, 1½ to 2 meters long, 1 meter deep and wide enough to work in, should be dug at right angles to the direction of the pull. The side of the trench towards the strain should be slanted so that it is at right angles to the direction of the pull. Another trench about 3/10 of a meter (12 inches) wide is dug, so that it intersects the main trench at a 90° angle in the middle. The bottom of this trench should be parallel to the strain and should meet the bottom of the main trench. A log 14 to 15 centimeters (5½ to 6 in.) in diameter is normally used for the deadman. The log is then put into the main trench and covered with dirt, with the exception of the part adjoining the second trench. If the dirt is not firm, stakes, the same length as the depth of the trench, should be placed between the deadman and the sloping side in an upright position. The standing end of the rope from the installation is passed directly around the deadman, so that any turning action of the log will tend to dig it in deeper (fig. 55). *The picket holdfast* is easier to construct but will not hold as much strain as the deadman. Stout pickets 6 to 7 centimeters (2.3 inches) in diameter are driven at least 1 meter into the ground, one behind the other with ½ to 1 meter remaining above the ground. Starting with the forward picket, secure the head of each by lashing it to the base of the next one. The lashings should be as tight as possible. The pickets should be driven at right angles to the strain and the distance between them should be several times the height of the pickets above the ground. The anchor rope is tied at the base of the closest picket. On the *piton anchor* at least four pitons are driven, with care taken in checking firmness and the solidness of the rock. Snaplinks are placed in the pitons and the end of the climbing rope is brought through the snaplinks until enough has been pulled through to make the proper tieoff (3 to 4¼ meters). A three-loop bowline is tied in the climbing rope, near the first piton, in what is to be the stake line. A snaplink is placed in each of the three loops of the bowline. A bight is brought down from the rope between each of the pitons. Each bight is snapped into one of the loops of the bowline. The running end is then tied on to the static line in front of the bowline with an anchor knot, in such a manner that it will slide up against the three-loop bowline. If it is not possible to have enough static for the span with this system, sling ropes may be used through the snaplinks on the pitons; the static line is tied to the sling rope with an anchor knot, thereby leaving all of the static line for the span. If there is a shortage of snaplinks, they can be omitted from the three-loop bowline and the rope threaded through the loops of the three-loop bowline.

Tightening Knots. For tightening fixed ropes, suspension traverse or any other similar installation, the following can be used:

Transport knot. A butterfly knot is tied in the static line far enough in front of the anchor to allow for tightening of the rope, with the bight of the butterfly approximately ⅓ meter long. This knot should be placed so that it acts as a safety for the load descending (approximately 1.2 meters from the lower anchor when tightened) preventing it from hitting the lower anchor. A pulley effect, for tightening the static line, is obtained by inserting a snaplink into the butterfly, passing the running end around the anchor and back through the snaplink in the butterfly. When the static line has been made sufficiently taut, it is tied off at the snaplink with a transport knot.

Prussik tightening knot. A butterfly knot is tied in the static line far enough in front of the anchor to allow for tightening of the rope with the bight of the butterfly approximately ⅓ meter long. This knot should also be placed so that it acts as a safety stop for the load descending, preventing it from hitting the lower anchor. A pulley effect, for tightening the static line is obtained by inserting a snaplink into the upper loop of the butterfly, passing the

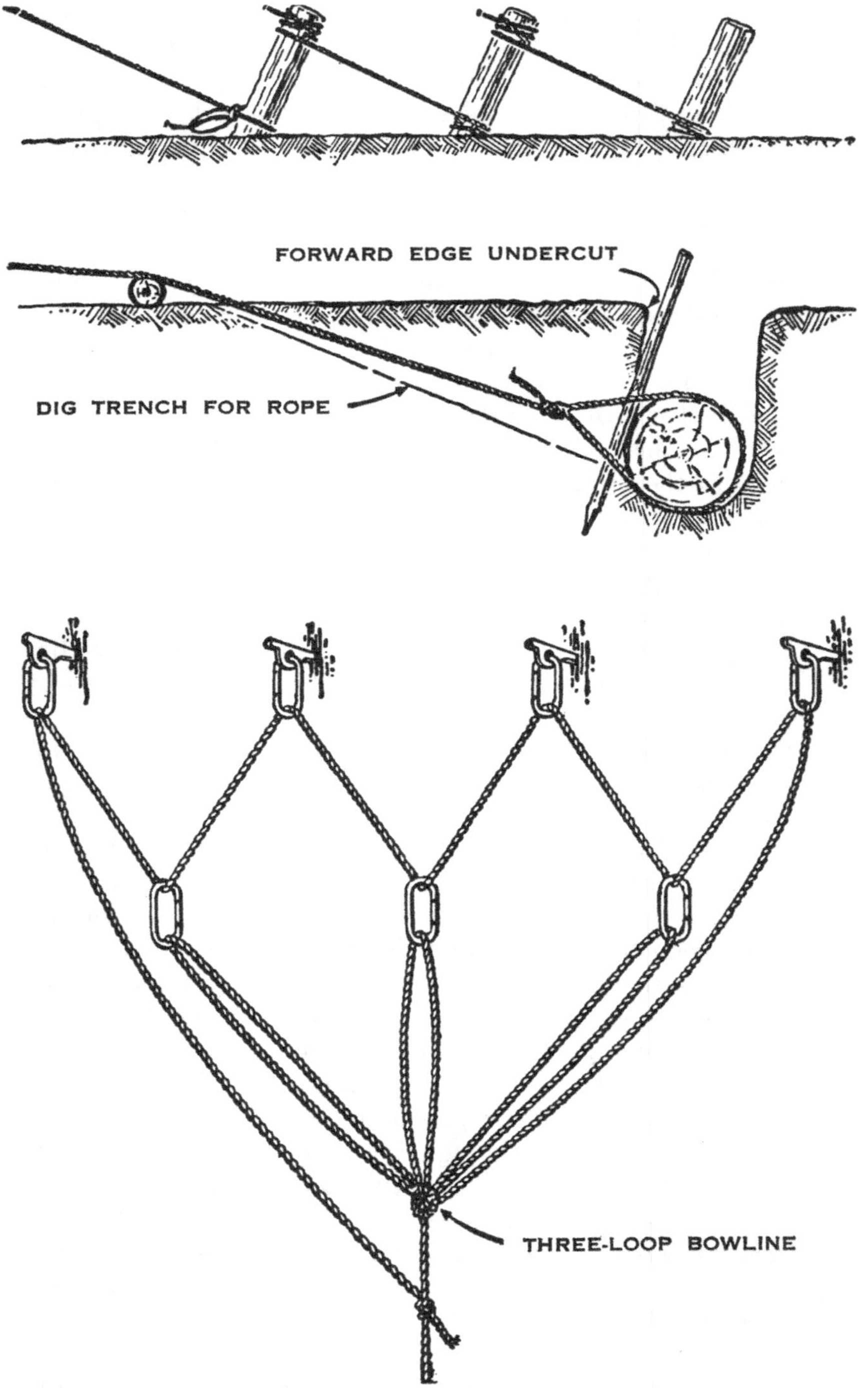

Figure 55. Artificial anchor-picket holdfast, deadman (timber in trench), pitons.

running end around the anchor, through the prussik knot which is made with the bight of the butterfly, and finally through the snaplink. The prussik knot, as the static line is tightened through use of mechanical advantage, acts to cinch the tightening rope. The running end is then secured with a half hitch on the rope opposite the prussik knot.

A-*frame*. An A-frame is constructed in the following manner:

Two sling ropes are tied together and secured to one of two sturdy ½ to 3 meter poles (approximately 7½ centimeters in diameter), with a clove hitch ½ meter from the top leaving a short running end of approximately ⅓ meter. The rope is wrapped horizontally around the poles six to eight times and vertically four times. Finally, the ends are tied tightly with a square knot. When the bottoms of the poles are spread apart, the resulting bipod forms the A-frame. If there is any danger of the bottom of the A-frame spreading, the poles should be braced by tying a sling rope between the two poles, tying it on each leg with a clove hitch. The feet should be dug in whenever possible or sandbagged to prevent slipping forward or backward.

Simple fixed rope. A simple fixed rope is made by anchoring one end of a climbing rope and using the line formed to aid in climbing. The procedure is as follows:

No. 1 ties in, plans route, and climbs on signal from No. 2 who belays or sees that the rope runs free. At top of pitch, No. 1 takes up slack and either ties the rope to an anchor point, or gets into a body belay. After installation, the unskilled men then climb, using the rope for all desired aid. Only one man climbs at a time, and signals to the next man, "climb", when he has reached the top. The last man up retrieves the rope and coils it.

Section III. ROPE BRIDGES

TIME ALLOTTED:	2 hours
TOOLS, EQUIPMENT, AND MATERIALS:	(Per 6–8 Soldiers) 4 climbing ropes, 4 snaplinks, 4 sling ropes
UNIFORM AND EQUIPMENT:	Field uniform w/sling rope, snaplink and heavy leather gloves
REFERENCES:	FM 31–72 and TM 5–279
TROOP REQUIREMENTS:	Medical aid man with litter jeep

Note. Three-rope bridges are covered by conference, inspection, and use of fixed installation only.

Rope bridges provide temporary and improvised systems for crossing streams, small rivers, gorges, etc., where the span is too great, the traffic not too heavy, and where there would be a saving in time over crossing by other methods or locating and using a bypass.

One-Rope Bridge. The one-rope bridge is constructed of a nylon rope or a Manila rope and anchored with a round turn and two half hitches and tied off with a transport tightening knot. It can be crossed by utilizing either of the following methods:

1. British crawl—Crosser lies on top of the rope with left instep hooked on rope (left knee bent) and right leg hanging straight for balance. After assuming this position, the crosser pulls himself across the rope with hands and arms.
2. Sling seat (Tyrolean traverse)—Crosser ties a rappel seat and by means of a snaplink hooks himself to the rope. Crosser is then hanging by his hands and a snaplink. He then pulls himself across with hands and arms while keeping his legs together and back slightly arched.

Note. Nylon rope may be used on crossing up to 18⅓ meters. For longer spans Manila rope should be used. This is due to the greater stretch factor in nylon rope.

Two-Rope Bridge. Constructed in the same manner as the one-rope bridge except that two ropes are used, one above the other and are spaced approximately 1.2 to 1.8 meters apart at the anchor points. Spreaders may be used to prevent excessive spreading between the two ropes. Spreaders are constructed of nylon sling ropes, placed perpendicular to the two ropes and tied off at each end with a round turn and two half hitches. Spreaders are placed approximately 4.5 meters apart. Construction is accomplished by tying a sling rope carrier and utilizing a rappel seat. Man constructing the bridge can sit in the rappel seat and by sitting on the bottom rope accomplish the tying of the spreaders.

Three-Rope Bridge (fig. 56). This bridge is established as a more permanent installation and for larger volume of traffic than the two-rope bridge. Spans of up to 30 meters may be set up using the nylon climbing rope. For spans in excess of 30 meters, Manila or sisal rope must be used.

1. The bridge consists of three main lines known as the handrails and tread rope. Additional ropes used as suspenders are sling ropes of from ¼-inch diameter up to ½-inch diameter depending on type most readily available.
2. Once an area with suitable anchors has been located, the two handrails and the tread rope are laid out parallel to one another approximately 1 meter apart with the tread rope in the center. Allow enough rope on each end to accomplish a tie-off at the anchors. Begin at one end and pace off the tread rope

with paces just short of a normal pace. At each pacemark lay out a sling rope so the center is across the tread rope. Tie the sling rope to the tread rope using a clove hitch, so the ends come out of the knot on the *bottom* side of the tread rope. The ends of the sling rope are tied to the handrails with a round turn and two half hitches so that the handrail comes up to the elbow of the average man. (Approximately 1 meter.) This procedure is continued the length of the span until all suspenders are tied in. The bridge thus prepared is transported to the bridge site and moved across the area to be spanned on a construction line previously put in. The construction line is a single strand anchored and tightened the same as a single rope bridge. The three-rope bridge is fastened to the construction line by snaplinks and a hauling line attached to the end of the bridge. As the bridge is pulled across with the hauling line, additional snaplinks are snapped onto the bridge and onto the construction line to prevent the bridge from sagging or snagging. The bridge is tightened as a two-rope bridge with emphasis on getting the same amount of tension on each of the handrails and the tread rope. If equal tension is not obtained, the spreaders will pull crooked and require major adjustment.

3. Adjustment of the bridge is necessary to insure stability and ease of crossing. The handrails must be forced apart 1 to 1.2 meters and fastened out to enable personnel to walk across. Two methods are used; one employing a spreader bar 1.5 to 2 meters in length and approximately 4 centimeters (1½-inches) in diameter, the other employing ropes. The bar is

Figure 56. Three-rope bridge.

tied to the handrails at each end of the span with a clove hitch around the bar. The second method is to tie separate ropes to the handrails using an "end of the rope" prussik knot and then tighten the ropes off at as near a 90° angle as possible. This second method requires additional anchors near the ends of the bridge. Additional stabilizing lines can be tied to the tread rope at the center and tightened to reduce sway and bounce.

4. To cross the bridge, both hands are slid along the handrails keeping contact with the ropes at all times. The feet are placed on the tread rope at the point where the clove hitch is located so that the toe points to the *outside*, the heel to the *inside* over the tread rope, and the instep is directly on the tread rope and clove hitch of the spreader rope. By so stepping there is little danger of the foot slipping off to one side or the other of the tread rope. A slow steady pace is maintained in crossing so that a minimum of bounce develops. Should the bridge begin to swing and bounce, all persons should stop where they are and push out on the handrails until the bridge steadies. The interval between men with full packs and weapons should be kept at no less than 6 meters.

Section IV. SUSPENSION TRAVERSE

TIME ALLOTTED:	2 hours
TOOLS, EQUIPMENT, AND MATERIALS:	3 climbing ropes, 4 snaplinks, sling rope
SOLDIER UNIFORM AND EQUIPMENT:	Field uniform w/sling rope, snaplink, and heavy leather gloves
REFERENCE:	FM 31–72
TROOP REQUIREMENTS:	Medical aidman w/litter jeep.

A suspension traverse is a rope installation used to bridge ravines, rivers, streams, or cliffs. It can be constructed both vertically, to raise or lower personnel or cargo from one level to another, and horizontally.

1. Materials required for establishing a suspension traverse are—
   a. Two 120-foot climbing ropes for static line (span up to 100 feet). One ¾-inch or 1-inch Manila rope for static line. (Span over 100 feet.)
   b. Sufficient climbing rope or ½-inch Manila rope for belay line.

c. Two snaplinks for transport tightening knots when double rope is used for the static line, or a single pulley, if available, for use with Manila rope. (Double snaplink pulley is adequate for use with Manila also.)

d. Two snaplinks for static line carrier.

e. One sling rope for carrier affixed to static line.

2. Characteristics of a good installation location:

a. Good anchor points.

b. Good loading and unloading platforms.

c. Cover and concealment from enemy observation and flat-trajectory weapons.

3. Construction:

a. The static line(s) is tied off at the higher end of the installation with an anchor knot. (A round turn and two half hitches is preferred.)

b. The lower end of the traverse static line is tied off at the bottom with a transport tightening knot. This knot is always tied at the lower end, thus serving as a safety stop should the cargo break loose from the belay line.

c. A carrier is made by tying the ends of a sling rope together with a square knot and two half hitches. The knot is then offset on the double rope ⅓ of the way down. An overhand knot is tied immediately above the square knot. A second overhand knot is tied just below the square knot. This leaves one-half of the doubled sling rope with no knots on it. This enables the operators to leave the knotted side of the carrier fastened to the snaplinks on the static line while the clean side is passed through the sling seat or cargo lashing.

d. The two snaplinks fastened on the static line are positioned so that one gate opens to the right and down while the other opens to the left and down. This is an additional safety factor to prevent a load from falling should a foreign object strike the snaplinks from either side.

e. The belay rope is tied to the carrier on the solid strand of rope opposite the square knot in the middle loop. It is tied using a round turn and a bowline. Should more than one rope be necessary, the two ropes are tied together with a square knot and two half hitches or a double sheet bend.

f. Where the belay line is apt to snag or hang on brush or rock between anchor points, it is tied up to the static line by placing snaplinks every 9 or 18 meters. The snaplinks are snapped into butterfly knots tied into the belay line. As the snaplinks come up the static line they are unsnapped from the static line but left on the belay line ready for snapping back on as a cargo is lowered.

g. Where the suspension traverse is horizontal, two belay or hauling lines are affixed to the carrier for movement of cargo in either direction.

4. Should it be necessary to raise the static line in order to load or unload cargo at either end of the traverse, an **A**-frame must be anchored to the static line to prevent its slipping out from under the static line. This is done by tying a close hitch with the center of a sling rope on one pole of the **A**-frame above the static line. With the ends of the sling rope, "end of the rope", prussik knots are tied on the static line so that one knot is on each side of the **A**-frame.
5. A maintenance crew must stay with the installation to constantly check on tightness of knots and correctness of the installation. Periodically, it may be necessary to loosen the static line to prevent undue fatigue. At other times it will be necessary to tighten the static line because of stretch that has developed. When Manila or sisal rope is used, a 5 percent slack should be allowed to avoid undue stress being put on the rope.

Section V. VERTICAL HAULING LINES

TIME ALLOTTED:	2 hours.
TOOLS, EQUIPMENT, AND MATERIALS:	(Per 6–8 Soldiers) 3 climbing ropes, 2 poles 8 feet to 10 feet long (4-inches diameter), 4 snaplinks.
SOLDIER EQUIPMENT AND UNIFORM:	Field uniform w/sling rope, snaplink, and heavy leather gloves.
REFERENCE:	FM 31–72.
TROOP REQUIREMENTS:	Medical aidman w/litter jeep.

A vertical hauling line is an installation for moving equipment or men up vertical or near vertical pitches. It is often used in conjunction with a fixed rope where the fixed rope is used for troop movement and the hauling line for equipment. It can also be superimposed over the fixed rope in difficult places as an additional aid for troops. Generally, three climbing ropes, four (4) snaplinks and the necessary equipment for the construction of an **A**-frame are needed for this installation; but any expedients may be used that will aid the construction.

To install, select a route which has good top anchor points, natural loading and unloading platforms and which affords sufficient clearances for easy hauling of equipment or troops. The ideal platform at the top will allow construction of the vertical hauling lines without the use of an A-frame. In such cases, anchors (trees or rock nubbins) close to the edge and high enough above the unloading platform to allow easy clearance are located, thereby cutting down on installation equipment and time. Otherwise, the construction is as follows: Construct an A-frame. The anchor rope is doubled and the end with the bight placed between the top V of the A-frame so that one-foot long bight hangs down. A clove hitch is made on each side of the bight a foot from the closed end. One of these is placed on the left pole of the upper part of the A-frame and the other on the right pole. The long part of the rope is then secured to separate anchors, if available, with a bowline knot. When weight is applied to the bight in the A-frame, the whole frame will lean forward and allow for clearance over the lip of the cliff when loads are hauled up. Another rope is then tied to a separate anchor (or the same anchor if none other exists) passed through the A-frame, and the remaining portion is knotted with overhand knots evenly spaced 20 to 25 centimeters (8 to 10 inches) apart. This is the rope used by the troops as a simple fixed rope as they are being hauled up the pitch. Two snaplinks are inserted in the bight hanging from the A-frame to form a pulley. One snaplink opens in, while the other opens out. A pulley rope is formed by tying the ends of the climbing rope with a square knot and half hitches and inserting it into the snaplinks. Run the completed knot up against the snaplink pulley and then form a belt below it by tying two butterfly knots just far enough apart so they can be snapped together with a snaplink to form a belt. On the opposite strand of the pulley rope, the same procedure is followed except that the belt is formed just below the point where the rope touches the ground at the bottom of the installation. There are several ways of tying into the pulley rope: with a sling rope seat snapped into the top butterfly knot of either belt; with the butterfly knot belt; or with a sling that can be worn under the armpits. If only two climbing ropes are available, the anchor rope and the knotted rope can be combined, and a double sling rope used for a pulley loop.

If equipment only is being hauled up, it is not necessary to have a knotted rope, but it may be necessary to use a belay rope as a guide line on the loads from the bottom to prevent damage from striking the cliff. To move materials or troops up one side of the hauling line, the other side is pulled from below. Troops using the hauling line for movement must employ all applicable principles of climbing.

Section VI. FIXED ROPES

TIME ALLOTTED:	2 hours.
TOOLS, EQUIPMENT, AND MATERIALS:	(Per 6–8 Soldiers) 2 climbing ropes, pitons, snaplinks, sling ropes, piton hammer.
INSTRUCTIONAL AIDS:	Several posts for anchors or wooded area.
SOLDIER UNIFORM AND EQUIPMENT:	Field uniform w/sling rope, snaplink and heavy leather gloves.
REFERENCE:	FM 31–72.

Fixed ropes are installed by Soldier assault climbing teams and used to assist trained men with heavy loads over difficult and varied terrain, safely and quickly. It is also used as a guideline, especially at night. This installation differs from the simple fixed ropes in that it employs many anchor points and is of permanent nature. To install, No. 1 ties into the leading end of the rope and moves up the selected route securing the rope temporarily to snaplinks inserted in pitons he has driven, or by means, of sling ropes tied to natural anchors. No. 2 remains at the starting point, paying out the rope and seeing that it runs freely while belaying No. 1 if he must climb over difficult terrain. He warns him when only twenty feet of rope remain. No. 1 locates an anchor point and anchors the upper end of the rope. If more rope is required, the No. 2 man comes up the first section and brings the necessary rope and equipment. Each additional section is put in as described above. When No. 1 reaches the top anchor point, he secures the end of the rope and moves down, tying the sling ropes to the fixed ropes with knots and securing the sling ropes to the anchors, or securing the fixed ropes to snaplinks in pitons with a figure 8 slipknot. He is assisted by No. 2. In this manner each section between anchors is tightened independently. The prussik knot in sling ropes and slipknots in the snaplinks serve to both increase the tension of the rope, and to make each section independent of any other section. This is a safety factor to insure that in the event of a break in one section of the rope other sections will not be affected. The prussik knot should be tied slightly above the anchor point. The figure 8 slipknot should be at least 12 inches from its anchor point after tightening. Each climbing rope in the fixed rope, as it is reached, is tied to the next section with a square knot and half hitches, and is not tightened separately. The transport knot or prussik tightening knot is used for the lower anchoring knot. More anchor points are needed on difficult terrain than are necessary when the going is relatively easy and the rope is used as a guideline. Many times the necessary equipment, such as sufficient sling rope for anchor points, will not be available. It is possible to anchor to many points in such cases with the climbing rope alone. In tightening down, a butterfly can be tied in the fixed rope above the anchor point (tree or rock nubbin). The rope is brought around from the side opposite the route, a bight of it passed through the butterfly, the rope tightened, and tied off with a transport knot. The rope then continues to the next anchor point. This method, of course, requires additional climbing rope to be used as slings. When snaplinks are not available and you have plenty of sling ropes, they may be used in the pitons or tied directly to an anchor. Maintenance of the fixed rope is the responsibility of the team that installed it. It is their duty to remove loose rocks, brush, and any other hindrance to the rope and troops. They must protect the rope with padding where necessary. They must also see that there are enough guides to help the troops in difficult places and control traffic, e.g., eliminate crowding and/or prevent large gaps from developing between anchor points. At times, where the anchors are far apart (easier terrain) and while moving in darkness, it is necessary to see the man ahead in order to keep the rope functioning smoothly. A diagonal or traversing route should be chosen for reasons of safety. The route should not be so difficult that the troops arrive on top exhausted, yet it should be steep enough to guide troops up without undue delay or distance traveled. Troops moving over a fixed rope must remember and use all the techniques of mountain walking and climbing. Use the fixed rope for direct aid only when necessary, but always have at least one hand on the rope. On difficult pitches employ friction footholds by leaning back and climbing hand-over-hand on the rope. The fixed rope is also used as an aid in descent. Variations of the above system are used often. The rope may be anchored directly at the bottom, the No. 1 laying out the rope and selecting anchor points; the No. 2 man follows him tightening the rope. The tightening knot is put in at the top. The reverse of this system is also very useful, starting at the top and working down. If working in a military climbing team, the balance of the men act as a security.

Section VII. MOUNTAIN WALKING

TIME ALLOTTED:	40 minutes
PERSONNEL:	Principal instructor; 5 assistant instructors.
REFERENCE:	FM 31–72
SOLDIER UNIFORM AND EQUIPMENT:	Field uniform w/sling rope, snaplink, and heavy leather gloves.

Note. During the body of the conference, demonstration is taking place. The demonstrator performs before the class each fundamental and technique as it is discussed.

Mountain walking is divided into four different techniques dependent on the general formation of the ground to be overcome. Included in all these techniques are certain fundamentals which must be mastered in order to obtain a minimum expenditure of energy and loss of time. These are—that the weight of the body must be kept directly over the feet and the sole of the shoe must be placed flat on the ground. This is most easily accomplished by taking short steps and a slow steady pace.

Walking on Hard Ground. Hard ground is generally considered to be firmly packed dirt which will not give away under the weight of a man's step. When ascending, the above mentioned fundamentals should be applied with the following additions. The knees must be locked on every step in order to rest the muscles of the legs. Steep slopes must be traversed and, if necessary, climbed in a zigzag manner rather than straight up. Turning at the end of each traverse should be done by stepping off in the new direction with the uphill foot. This prevents crossing of the feet and possible loss of balance. In traversing, the full sole principle is most easily accomplished by rolling the ankle away from the hill at each step. For narrow stretches the herringbone step may be used; that is, ascending straight up a slope with the toes pointed out and using all the other principles mentioned so far. Descending is most easily done by coming straight down on a slope without traversing. The back must be kept straight and the knees bent in such a manner that they take up the shock of each step. Again it must be remembered that the weight has to be directly over the feet, and that the full sole must be placed on the ground at every step. Walking with a slight forward lean and with feet in a normal position will make the descent easier.

Grassy Slopes. In mountainous terrain, grassy slopes will usually be made up of small hummocks of growth rather than one continuous field. Therefore, in ascending it will be found that while the techniques previously mentioned are applicable, it is better to step on the upper side of each hummock where the ground is more level and secure than on the lower side. Descent is best accomplished by traversing; using the platform on the upper side of the hummock to place the foot on. If running it may be best to use the traverse with a hopskip. The hopskip is a hopping motion in which the lower foot takes all the weight and the upper foot is used for balance only. The hopskip is also useful on hard ground and scree when descending.

Scree Slopes. Scree slopes consist of small rocks and gravel which have collected under cliffs. The size of the scree varies from sand to pieces about the size of a man's fist. Occasionally it occurs in mixtures of all sizes but normally scree slopes will be made up of the same size particles. Ascending scree is extremely difficult and should be avoided whenever possible. All principles of ascending hard ground apply, but each step must be packed carefully so that the foot will not slide when weight is placed on it. This is best accomplished by kicking forward and in, with the uphill side of both the upper and lower foot, using a full sole on the platform made. Coming down in a straight line again is the best way to descend scree. Here it is important to keep the feet pointed straight down, as well as keeping the back straight and the knees bent. Since there is a tendency to run down scree, care must be taken that too great a speed is not attained and control lost. By leaning slightly forward, greater control can be obtained. When a scree slope must be traversed and no gain or loss of altitude is desired, the hopskip may be used.

Talus Slopes. Talus slopes are similar in makeup to scree slopes, with the exception that they are composed of locked rocks of larger size resting upon each other in various degrees of stability. The technique of walking on small talus is to step on top of and on the uphill side of the rocks. This prevents them from tilting and rolling downhill. All other fundamentals previously mentioned are applicable. On larger blocks of talus a rhythmic, balanced springing from rock to rock is employed keeping constantly alert as to the route ahead so that rhythm will not be lost.

General Precautions. It is of the utmost importance that rocks are not kicked loose in such a manner that they roll downhill. Falling rocks are extremely dangerous to the men below and make a great deal of noise. Carelessness by one man in this respect can cause the failure of a well-planned mission since one rock no bigger than a man's head can kill or severely injure several men and ruin all security measures. Stepping over rather than on top of obstacles such as rocks and fallen logs will help toward avoiding fatigue. Usually it will be found that talus is easier to ascend and traverse, while scree is a desirable avenue of descent. If there is danger of causing rocks to slide or roll while ascending or descending, it may be better to make a long continuous traverse. As a general rule, one should not jump in the mountains. Areas on which to land are usually small, uneven and have loose rock or dirt, which make landings precarious. A man with even a slightly sprained ankle is extremely handicapped in rough terrain.

Section VIII. BALANCE CLIMBING

TIME ALLOTTED:	1 hour.
TOOLS, EQUIPMENT, AND MATERIALS:	Climbing rope per 3 Soldiers.
PERSONNEL:	Principal Instructors; 1 assistant instructor per 6 Soldiers.
REFERENCE:	FM 31–72.
SOLDIER UNIFORM AND EQUIPMENT:	Field uniform w/sling rope, snaplink and heavy leather gloves.
TROOP REQUIREMENTS:	Medical aidman w/litter jeep.
TRANSPORTATION:	One 2½-T truck per 18 Soldiers, one (1) ¼-T truck.

Note. A suitable rock area not too steep with minimum exposure should be used for initial instruction.

Balance climbing is the type of movement used to travel on steep slopes. It is a combination of the balanced movement of the tightrope walker and of a man ascending a tree or ladder.

1. *Body position.* The Soldier must climb with the body in balance, which means that the weight should be poised over the feet or just ahead of them as he moves. The feet, not the hands, carry the weight whenever possible. The hands are used primarily for balance. Feet will not hold well when the climber leans in toward the rock. With the body in balance the climber moves with a slow rhythmic motion. Three points of suspension, two feet and one hand for example, are used whenever possible. Handholds that are waist to shoulder high are preferable. Relaxation is necessary because tensed muscles tire quickly. When resting, the arms are kept low where circulation is not impaired. Use of small intermediate holds is preferable to stretching and clinging to widely separate big holds. A spread-eagle position, in which a man stretches so far he cannot let go, should be avoided. In descents, the climber faces out where the going is easy, sidewise where it is more difficult, and faces in where it is most difficult. He uses the lowest possible handholds.

2. *Types of holds.*

   a. Pull holds. These are holds pulled down upon and are the easiest to use. They are also the most likely to break out.

   b. Push holds. These are holds pushed down upon to help the climber keep his arms desirably low and rarely break out, but are more difficult to hold on to in case of a slip. A push hold is often used to advantage in combination with a pull hold.

c. Friction holds. These are holds, dependent solely on the friction of hands or feet against a smooth surface. They are difficult to use because they give a feeling of insecurity which the climber tries to correct by leaning close to the rock, thereby actually decreasing his security. They often serve well as intermediate holds, giving needed support while the climber moves over them, but would not hold him were he to slip.

d. Jam holds. These holds involve jamming any part of the body or extremity into a crack. This can be done by putting the hand into the crack and clinching it into a fist or by putting the arm into the crack and twisting the elbow against one side and the hand against the other side. The same can be done with the boot, knee and leg.

e. Pinch holds. These holds are those used where pinching action of the fingers and thumb can be employed on a projection or ledge of rock. When just the tips of the fingers and thumb are used the hold is used only for balance. However, when the entire hand can be used to pinch, the hold may be used for a direct assist in moving the body. A variation of the pinch hold is the use of both hands with palms pressed towards each other with a narrow rock projection in between.

f. Variations. The holds previously mentioned are considered basic and from these any number of combinations and variations can be used. The number of these variations depends only on the limit of the individual's imagination. There are a few which require special techniques—for example—

(1) *Lie back.* This is done by leaning to one side of an offset crack with the hands pulling and the feet pushing against the offset side.

(2) *Chimney.* Cross pressure is exerted between the back and the feet, back and knees, etc.

(3) *Inverted pull or push hold.* These are sometimes underholds whereby cross pressure is exerted between hands and feet.

(4) *Shoulder stand.* The shoulder stand, or human ladder, is used to overcome a holdless lower section of pitch in order to reach the easier climbing above. Lower man is anchored to rock and belays leader.

g. Footholds. Footholds are categorized as: Step-push, Jam and Friction. The service shoes with rubber sole will hold on steep slab. On steep slopes the body should be kept vertical with use being made of small irregularities in the slope to aid friction. Footholds ½-inch wide can be sufficient for intermediate holds, even when they slope out. It is better to use the side of the shoe, rather than the toe as it is much stronger, less flexible and less tiring. As much of the sole as possible should be used against the rock.

h. Use of holds. The use of holds is just as important as the holds themselves. A hold need not be large to be good nor need it be solid, so long as the pressure applied holds it in place. The climber must roll feet and hands over his holds, not try to skip or jump from one to another. It is, however, often desirable, while traversing, to use the hop-skip, in which the climber changes feet on a small hold so that he may move sideways more easily. A slight upward hop followed by precise footwork will accomplish this useful step. Test all holds before using. NEVER LUNGE FOR A HOLD.

3. *Margin of safety.*

a. A margin of safety is the protective buffer a climber keeps between what he knows to be the limit of his ability and what he actually tries to climb.

b. The climber learns his margin of safety by climbing close to the ground, or tied to a rope held or paid out by a trained man above. He climbs first on the easy and obvious holds, next on to the more difficult ones, and finally on difficult pitches until he reaches the limit of his ability.

c. The margin of safety should be calculated not only for the pitch immediately ahead but for the entire climb. The climber should plan his route and movement far enough ahead so that he never finds himself in difficulties beyond his ability. The leader of a group must allow for the limitations of his men.

d. Before beginning any climb on rock, all loose mud, dirt, sand, and foreign matter should be removed from the bottoms of the boots. Even a small amount of debris can cause a fatal slip. This can usually be accomplished by kicking the side of the boot against a rock.

e. The climber should always climb rock formations with bare hands in order to deftly feel each handhold.

f. The climber should remove from the front of his body all articles of clothing or equipment that might catch or push on rock projections and cause him to lose his balance. Even a slight push or pull, if unexpected, may cause a fatal fall.

g. Use of knees, elbows, or buttocks as a *primary* hold should be avoided since the climber has no direct muscular control over these parts for holding purposes. In addition, knees and elbows are easily bruised on rough rock surfaces. However, use of knees, elbows, and buttocks in conjunction with other hold variations is permissible and often the best solution to a climb.

Section IX. BELAYS

TIME ALLOTTED:	One and one-half hours
TOOLS, EQUIPMENT, AND MATERIALS:	(Per 6–8 Soldiers) 2 climbing ropes, 2 anchor points, 2 pitons, 2 belaying logs (75–90 pounds each)
PERSONNEL:	Principal instructor; 1 assistant instructor per 6 to 8 Soldiers
REFERENCE:	FM 31–72
SOLDIER UNIFORM AND EQUIPMENT:	Field uniform w/sling rope, snaplink and heavy leather gloves
TROOP REQUIREMENTS:	Medical aidman w/litter jeep

Note. Tie-ins for party climbing and rescue work are taught in conjunction with belays. Initial instruction is taught on steep slopes before progressing to cliff area.

In party climbing two or three men are tied-in to a 120 foot length of rope. Belaying provides the necessary safety factor, enabling the leader to climb. Without belaying skill, the use of rope in party climbing is a hazard, not a help. When one man is climbing, he is belayed from above or below by another man, who may use any one of several belay positions. Belaying is also used to control descent on fixed installations.

Procedure for all Positions. The belayer must perform the following duties:

a. Run the rope through his guiding hand, which is the hand on the rope running to the climber, and around the body, or other anchor to his braking hand, making certain that it will slide readily.

b. Anchor himself to the rock with a portion of the climbing rope, or his sling rope if his position is unsteady.

c. Make sure remainder of rope is laid out so as to run freely through the braking hand.

d. See that rope does not run over sharp edges of rock.

e. Avoid letting too much slack develop in the rope through constant use of the guiding hand, except where this hand is used as a brace. Be gentle, and tug the line running to the climber, thus sensing his movement. Avoid taking up slack too suddenly, as this may throw the climber off balance.

f. Brace well for the expected direction of a fall, so that the force will, whenever possible, pull the belay man more firmly into position. A climber should not trust a belay position which he has not tested.

g. Where necessary, seek a belay position that offers cover and concealment.

h. Be able, in case of a fall, to perform the following movements automatically.

(1) Relax the guiding hand.

(2) Let the rope slide enough so that braking action is applied gradually. This is done by bringing hand slowly across the chest or in front.

(3) Hold belay position, even if this means letting the rope slide several feet.

Sitting Belay. This is the preferred position. Belayer sits and attempts to get good triangular bracing between his legs and buttocks. Whenever possible, legs should be straight, knees locked. The rope should run around the hips. If the belay spot chosen is back from the cliff edge, friction of rope over rock will be greater, and will simplify holding of falls, but the direction of the pull on the belayer will be directly outward.

Standing Hip Belay. This is a weaker belay position and is used only where it is not possible to use the sitting belay. An anchor is essential. A very strong well-braced position is assumed with one leg braced in the direction of a possible fall. If feasible, brace the back or shoulder against the wall. The rope from the climber comes up the leg braced in direction of fall, around the back just above the heavy part of the hips and around in front to the braking hand. If possible, it should be above the rope anchoring the belayer to the rock.

Anchored Belays. In this type of belay, the belayer anchors himself by use of the climbing rope or his sling rope. The anchor is usually a piton placed so that it will prevent the belayer from being pulled out of position. Rock projections and well-rooted trees may be used. Whenever possible, the belayer should always anchor himself for additional safety.

Piton Belay. As soon as the leader has placed a reliable piton, the direction of pull when he falls will be forward and up. The belayer should have a low position directly in line with this direction of pull, and should run the belay rope just below his buttocks. Both knees should be bent, to prevent the rope from sliding up and above the buttocks. The braking arm should be extended. When a fall occurs, the arm is brought in, with steadily increasing resistance, to a position in front of the body where as much rope as necessary is then allowed to slide through the hand to break the fall gradually. A fall is easier to hold by a piton belay than by a sitting or standing belay, especially if there are several pitons in place, because of the added friction between rope, rock, and snaplink. For this reason, it is essential that the belayer use a dynamic belay to prevent the fall from jerking to a sudden stop; likewise, he must not resist the fall too much when its impact first hits his body. The climber will signal "belay for piton", after he has snapped into his first piton.

Rock or Tree Belay. Where possible, the leader passes his rope behind rock projections or trees, which can serve the same protective purpose as a piton. He should avoid passing the rope over sharp edges, or crevices where the rope could jam or cause too much friction as he climbs beyond. When a rock or tree belay has been definitely established, the belayer should assume the piton belay position.

Rope Signals for Belays. After the belayer has found a belay position and has settled himself there he calls "on belay", "test", which is answered by the climber calling "off belay, testing". He then puts his weight gradually on the rope, until his full weight is held. Care must be taken not to jerk the rope suddenly when starting the test, as this might pull the belayer out of position. If position is satisfactory, the belayer calls "climb". The climber answers him with the signal "up rope", in order to have belayer take up the extra slack. As soon as the slack is up, the climber calls "climbing" and starts to climb. If during the testing, the belayer finds his position unsatisfactory, he must call, "off belay" and the climber must release all tension at once and reassume his belay position. The belayer must then find another position and repeat the test procedure. If this method is not followed, serious accidents may occur.

Static Belay. A static belay is where the belayer applies the braking action of locking the arm across the body with utmost speed, allowing the absolute minimum of rope to run through his hands. When applied, it creates sudden and sharp halts to the person or cargo being belayed.

Dynamic Belay. In the dynamic belay, the belayer allows rope to run through his hands as he gradually brings his braking hand across his body to produce braking action. The climber or cargo is brought to a gradual more gentle halt where less strain is put on the climber, the belayer, pitons, and rope.

Section X. RAPPELLING

TIME ALLOTTED:	4 hours.
TOOLS, EQUIPMENT, AND MATERIALS:	8 climbing ropes.
PERSONNEL:	Principal instructor, assistant instructor per 8 to 16 Soldiers.
INSTRUCTIONAL AIDS:	Log ramp approximately 25 feet high for initial instruction.
SOLDIER UNIFORM AND EQUIPMENT:	Field uniform w/sling rope, snaplinks and heavy leather gloves.
REFERENCE:	FM 31–72.
TROOP REQUIREMENTS:	Medical aidman w/litter jeep.

Note. **Initial instruction, demonstration and practical work is conducted on a slope before Soldier works on the log ramp. Advanced instruction and practical work is performed on progressively higher cliffs.**

Purpose. The climber with a rope can descend quickly by means of a rappel. This means sliding down a rope which has been placed around such anchor points as a tree, projecting rock, or several pitons. This is done in such a manner that the rope can be retrieved from the bottom. Several techniques may be used.

Establishing a Rappel.

1. In selecting the route, the climber should be sure the rope reaches the bottom or a place from which further rappels will reach the bottom.
2. The anchor point should be tested carefully, and inspected to see that the rope will run around it when one end is pulled from below.
3. If a sling rope must be used for a rappel point, it should have two turns around the anchor finishing off each turn with a square knot.
4. The first man down should—
   a. Choose a smooth route for the rope, free of sharp rocks.
   b. Place loose rocks, which the rope might later dislodge, far enough back on the ledge to be out of the way.
   c. Clear the route of brush and other obstructions.
   d. Prevent the doubled rope from twisting together by placing the index finger of the braking hand between the two ropes.
   e. See that the rope will run freely around the rappel anchor when pulled from below.
5. Each man down will signal "off rappel" by pulling alternately on each end of the rope, so that the rope runs across the rappel point. This will be barely audible at night and will also assure retrieving of the rope after everyone is down.

Figure 57. Ideally a smooth vertical surface is best for initial rappel instruction, but any vertical rock formation having a good access route and large base area is acceptable.

When silence is not necessary, the call "off rappel" should also be used.

6. When the last man is down, the rope is recovered. The climber should pull it smoothly, to prevent the riding end from whipping around the anchor point, and he should stand clear of falling rope, and the rocks which may be dislodged by it.
7. Inspect the rope frequently if a large number of men are rappelling on it.
8. Gloves should be worn on rappels.

The Body Rappel. The climber faces the anchor point and straddles the rope. Then he pulls the rope from behind and runs it around either hip and runs it diagonally across the chest and back over the opposite shoulder. From there the rope crosses, for example, the right hip to the left shoulder to the right hand. The climber should lead with the braking hand down and should face slightly sideways. He should keep the other hand on the rope above him just to guide himself and not to brake himself. He must lean out at a sharp angle to the rock, in order to provide his feet with the necessary friction to hold them onto the rock. He should keep his legs well spread and relatively straight for lateral stability and his back straight, to reduce unnecessary friction. The collar should be turned up to prevent rope burns on the neck. Gloves should be worn and any other article of clothing may be used as padding for the shoulder and buttocks.

The Hasty Rappel. Facing slightly sidewise from the anchor, the climber places the rope across his back. The hand nearest the anchor is his guiding hand and the lower hand is his breaking hand. To stop, the climber can bring his braking hand to the front of his body, locking it and at the same time turning to face up toward the anchor point. This rappel should be used only on moderate pitches or very short, steep pitches. Its main advantage is that it is easier and faster to use than the other methods, especially where the rope is wet.

The Seat Rappel.

1. *The Rappel Seat.* Place the sling rope across the small of the back so that the middle is above one hip. (This will cause one end to be shorter than the other in front.) Cross the ropes between the legs (front of the body toward the rear), then bring each end around the legs to the short side and tie off with a square knot and two half-hitches on one hip. Pull up on the cross ropes passing through the crotch, so that the center of the X then formed coincides with the horizontal rope coming across the lower part of the abdomen. Hold the snaplink, so the gate opens on the down side with the hook of the snaplink pointed toward the body. Snap down through the top of the X, include the horizontal rope and out through the bottom of the X. When the snaplink closes, rotate it one-half turn so that the gate is up and opens away from the body. *This is the only correct way to hook in with the snaplink!*
2. *Shoulder Method.* Pass the rope from the rappel point up through the snaplink, over one shoulder and back to the opposite hand (right shoulder to left hand). The same techniques are used in the descent as are used in the body rappel. This method is faster than the body rappel and cooler, except at the shoulder. It is safer for men with heavy packs, but requires more equipment and time for preparation.
3. *Hip Method.* The sling rope and snaplink are put on in the same manner as in the shoulder method. Stand on the side

of the rappel rope and snap it into the snaplink. TAKE SOME SLACK BETWEEN THE SNAPLINK AND THE ANCHOR POINT and pull it underneath, around then over the snaplink and snap it in again. (TAKE TWO TURNS WITH A SINGLE ROPE.) This will result in a turn of double rope around the solid shaft of the snaplink which does not cross itself when under tension. Facing the anchor, the climber descends using his upper hand as the guiding hand and his lower hand for braking. The rope should be grasped by the braking hand with the thumb pointing down above the hip. The braking is done by closing the hand and pressing the rope against the body in the small of the back. The friction generated heat necessitates the use of gloves for this rappel. The climber should lean at a sharp angle to the rock and make a smooth even descent. The feet should be held high and approximately shoulder width apart to provide stability. This rappel is fast and easy to use with heavy packs. However, if the rope is not snapped correctly into the snaplink, there is a good possibility that the gate will open and the rope will come out. The rope running through the snaplink in this rappel twists quite badly and at times is difficult to get out of. This would be especially disadvantageous at night. Loose clothing or equipment can work into the snaplink locking the rappel. Because of these reasons, the rappel must be checked before each descent.

Section XI. CLIFF EVACUATION

TIME ALLOTTED:	1 hour.
TOOLS, EQUIPMENT, AND MATERIALS:	Medical litter; 2 poles 2.4 to 3 meters (8 feet–10 feet) long, 2 bracing sticks 30 to 32 centimeters (12-inches–14-inches) long; 4 sling ropes; 3 climbing ropes.
PERSONNEL:	Principal instructor; 2 to 4 assistant instructors.
REFERENCE:	FM 31–72.
SOLDIER UNIFORM, AND EQUIPMENT:	Field uniform w/sling rope, snaplink, and heavy leather gloves.
TROOP REQUIREMENTS:	Medical aidman w/litter jeep.

Note. Techniques of lashing patient to litter and patient to rappeller is demonstrated away from cliff area. Conference and demonstration of the techniques involved are presented. Practical work can be integrated with other mountaineering instruction or as a part of a field patrolling exercise.

Preparation of Litter. To prepare a litter for the process of evacuation, place a stick, 30 centimeters (12-inches) long and at least 1-inch in diameter along each hinge joint and fasten it securely with wire or strong cord. This will prevent the hinge joint from collapsing. Place two poles, 2.4 to 3 meters (8 to 10 feet) long, on the bottoms of the litter stirrups, one on each side and running lengthwise to the litter. The larger end of the pole should extend midway between the stirrups, the end of the handles at the foot-end of the litter, and the smaller end well beyond the handles at the head-end. It is important that the ends be even with each other at the head-end, as they act as skids. These poles are also protection from the bottom of the litter. At the

Figure 58. Litter evacuation.

head-end of the litter, pass 2.4 to 3 meters (8 to 10 feet) of the end of a climbing rope through one litter stirrup, bringing it around and through again, forming a round turn. Two half-hitches are then put around the crossbar and brace, one on each side of the hinge joint. A round turn is made on the opposite stirrup, the end is then tied to the long part of the climbing rope with a bowline between the canvas of the litter and the upper ends of the handles. The rope acts as the belay rope on descent and to aid the raising of the litter on ascents.

The method of lashing a casualty to a litter described here, is a basic system and may be altered depending upon the type and location of the injury. Four 3.6 meters (12-feet) sling ropes are used. Two to lash the upper part of the body and two for the lower part. A team of two men is the most efficient, one man working on each side of the litter (fig. 58). The upper part of the body is lashed first. Two sling ropes are used, tied with a bowline on the upper part of the legs in the crotch. The crotch should be padded and the knot tied so that the rope does not bind. One sling rope on one leg and the other on the opposite leg. The ends of the rope are then brought diagonally across the body, under the arms, to the stirrup on the opposite side of the litter, passed through the stirrup and a round turn formed. They are then brought across the chest to the opposite stirrup, another round turn formed, the rope brought back on to the chest and tied off with a square knot. If the ropes are not long enough to reach across the chest and back again, they will be tied off on the chest after the first round turn on the stirrup. To tie the lower part of the body, sling ropes are tied to each of the upper stirrups with a round turn and two half-hitches or by bowlines, brought diagonally across the body to the stirrups on the opposite side of the litter at the foot, and round turns made on each stirrup. The knees of the casualty are bent slightly and the feet drawn back. The ropes are brought from the stirrups and wrapped around the feet bringing the rope across the bottom of the feet first, to act as a platform. Tie off with a square knot on the bottom or side of the feet, so that the knot will not form pressure on the arch of the foot.

Evacuation Procedures. As soon as the belay rope is tied to the litter, and before the patient is tied on, a man goes on belay to insure that the litter will not slide out of control. He remains on belay until the team leader calls for "off belay" on safely reaching the bottom of the cliff.

Two rappellers guide the litter down, one on each side of the litter. The rappellers, when necessary, pull away from the cliff on the lower handles to prevent the litter from catching or hanging on rock projections. One rappeller is designated to give all signals relative to "slack", "tension", and/or "up rope". Upon reaching the bottom

Figure 59. Piggyback evacuation.

of the slope, the litter and patient are immediately moved away from the slope to reduce danger from falling rock.

Piggyback Evacuation (fig. 59). Casualties who are not seriously injured, but cannot negotiate a descent by themselves, may be carried down a cliff by a carrier or rappeller. In this case, the casualty is belayed from above by means of climbing rope. The belay rope is tied around the chest of the patient with a bowline pulled snugly against his chest. The rappeller hooks into the rappel rope for a seat hip rappel. The casualty then straddles the carrier's back, positioning himself as high as he can. The center of a sling rope is placed under the patient's buttocks and passed to the front of the carrier. The ends of the rope are crossed on the carrier's chest, passed over his shoulders and under the armpits of the patient, to the rear where they are tied off with a square knot and two half-hitches. The rappeller puts his braking hand over the leg of the patient and grasps the rappelling rope, which passes under the patient's leg, in the same manner as for a seat hip rappel. If the patient is a particularly heavy man, an alternate braking method is used. The rappeller hooks up from the opposite side of the rappel rope, passes the trailing end of the rope around the small of the back and brakes on the same hip as he normally would. Before backing over the cliff, the rappeller ascertains that the belayer is "on belay" and prepared to control the gradual descent of the pair. The rappeller walks slowly down the cliff keeping his body well out in a normal rappel position.

APPENDIX VI

EXAMPLE SURVIVAL LESSON OUTLINE

I. Lesson Objective. To introduce survival techniques for the purpose of orienting and demonstrating to the Soldier certain sound and tried methods of survival in isolated areas or when forced by enemy action to isolate himself.

II. Teaching Points.

A—Survival is largely a mental outlook with the WILL TO SURVIVE the deciding factor.

B—Prompt action and a basic knowledge of first aid enables one to treat injuries and prevent disease.

C—Water is the basic requirement for survival. Without water, food is of little importance.

D—A knowledge of where, when, how to find, and how to prepare edible animal and plant life is essential to survival.

E—A knowledge of wild animal habits and individual ingenuity are necessary to procure wild life for food supply.

F—Shelters and fires will increase chances for survival while greatly reducing physical hardship.

III. Advance Assignment. None.

IV. Introduction.

A—*Gain attention.* Kill and dress wild animal (rabbit, squirrel, etc.). See figure 60.

B—*Orient Soldiers.*

1. *Lesson tie-in.* There are so many different emergency situations in a war that it is impossible to prescribe any definite survival formula. The best insurance for survival is the possession of basic knowledge.
2. *Motivation.* Either alone or in small groups, cut off behind enemy lines, lost in mountains or jungle, you may be required to live by your wits and nature alone for extended periods of time.
3. *Scope.* The first three periods of instruction will be devoted to an introduction to survival techniques. During the first period we will develop and discuss the "will to survive"; for the remainder of this period and in the second period you will visit____________________________ stations where

you will witness demonstrations of various survival techniques.

V. Presentation.

A—*First teaching point—survival.* Survival is largely a matter of mental outlook with the "will to survive" being the deciding factor.

Note. By using the letters in the word "survival," the instructors develop and discuss the "will to survive."

Figure 60. An outdoor rattlesnake feast.

1. Size up the situation.
2. Undue haste makes waste.
3. Remember where you are.
4. Vanquish fear and panic.
5. Improvise.
6. Value living.
7. Act like the natives.
8. Learn the basic skills.

Note. Class is divided into -------- groups and sent to indicated stations. Groups will rotate through the -------- stations.

B—*Second teaching point—first aid.* Prompt action and a basic knowledge of first aid enables one to prevent disease and injuries.

1. Accidents most common are—

a. Snakebite.

b. Severe lacerations.
c. Twisted and broken bones.
d. Eye injury.
e. Drowning or near drowning.

2. Medical aid will not be available when isolated.
3. Ready to treat with only:
   a. Commonsense.
   b. Knowledge of first aid.
   c. First aid packet.
4. Three "lifesavers."
   a. Stop bleeding.
   b. Protect wound.
   c. Treat for shock.
5. Personal hygiene is of paramount importance when isolated.
   a. Keep clean.
      (1) Daily washing of armpits, crotch, and feet.
      (2) Brush teeth regularly.
   b. Guard against cold injury. When exposed to severe cold, conserve your body heat by every means possible.
   c. Guard against injury caused by heat. Avoid strenuous exertion in the sun since heatstroke may result.
   d. Guard against insects and insectborne diseases. Every possible means should be used to avoid the contamination of food by flies and insect bites.
   e. Take care of your feet.
      (1) Do not wear dirty or sweaty socks.
      (2) Blisters are dangerous because they may cause fatal infections.
      (3) Some field expedients to protect feet are—
         (*a*) Scarf.
         (*b*) Bandage.
         (*c*) Gloves.
         (*d*) Handkerchief.

Note. Instructor demonstrates.

a. Use of snakebite kit.
b. Proper method of artificial respiration.
c. How to prepare field expedient splints and litters.

Note. Move to next station.

C. *Third teaching point—water*. Water is a basic requirement for survival. Without it food is of little importance. Under average conditions an individual needs at least a quart of water a day. A Soldier who knows how to use water intelligently will survive in reasonably good condition on a supply which, to another man, is insufficient and may cause death by thirst. If you are extremely thirsty, sip slowly and don't drink an excessive amount of water. Likewise, if you are hot from sun or from exercise, avoid drinking an excessive amount of water. If water is scarce and you are using a great amount of energy, you will lose less through perspiration by drinking small amounts at fairly frequent intervals, than by drinking a large amount at one time.

1. When looking for water, remember that the water table is usually close to the surface and can be reached by digging in low forested areas, along the seashore, and in flood plains of large rivers.
2. In all arid parts of the world there are numerous indications of the presence of water.
   a. Some plants grow only where ground water is close to the surface.
   b. In dry regions, dig where the sand is damp, in dry river beds or other low areas.
   c. Dew can be collected in useful quantities during a clear night.
   d. As a last resort, water may be obtained by breaking off a young desert tree at the base and removing the top.
3. Mountain snow on a clear day can be melted by placing a shallow container on a sunny exposure out of the wind. Dry mountain streambeds often contain water beneath the gravel stream bottom. Put your ear to the ground and listen for the trickle.
4. Sap is chiefly water and from many plants it is both fit to drink and readily available.
5. Many desert and other plants store water in their fleshy leaves or stems.
6. Water can be readily acquired from many type vines (grape vines), stems, and fruits.
7. It is often necessary to use muddy, stagnant, or polluted water. Water polluted by mud or animals is unpleasant but harmless if it is boiled.
   a. Water that merely has had the sediment eliminated is not purified. To be safe it must be boiled at least three minutes or longer. Halazone tablets will purify unboiled water. Let it stand for a half hour before drinking. If there is a slight chlorine smell, the water is safe to drink.
   b. Don't try to shortcut on water purification. Waterborne diseases are one of the worst hazards of tropical and subtropical climates. If you boil or chemically purify all drinking water, the dangers are reduced for contracting dysentery, cholera, typhoid fever, and parasitic infections.

Note. Show grapevine—water table.

D. *Fourth teaching point—food.*

1. Food follows water in the order of its importance in survival.
2. Food is all around you—growing, watching you, "right now." Wild food (plants, fruits, etc.) can be used to supplement rations, or you can live on it entirely.
3. Wild food you find will seldom be delicious and succulent because of the method of preparation, although you may make it more appetizing by proper cooking and adding available condiments. The tastes of food will often be strange and flat—possibly tough.
4. Remember as hunger increases, you become more and more like a basic animal; RIP, TEAR, GRUB, AND DIG.
   a. Plant foods. There are 300,000 kinds of plant food in the world; most are edible.
   b. Some tests for the poisonous ones are—
      (1) Burning, bitter or nauseating taste—milky sap.
      (2) Best test.
         (*a*) Boil for 5 to 20 minutes.
         (*b*) Taste—five minutes.
         (*c*) Swallow small portion.
         (*d*) Wait approximately 6 hours.
      (3) If animals eat the plant, you can also eat it.
   c. Types of plant foods to look for—
      (1) Tubers (potatoes).
      (2) Bulbs (onions).
      (3) Young stems and sprouts (asparagus).
      (4) Grain (wheat).
      (5) Nuts (pecan, hickory).
      (6) Fruits (berries).
      (7) Young and tender leaves (spinach).
      (8) Seeds (sunflower seeds).
   d. Preparation.
      (1) Boiling—change water.
      (2) Roast or bake.
   e. Animal foods.
      (1) Almost all animals are good to eat, including insects and reptiles. Anything with fur is good. Also mammals, bugs, fish, snakes, birds, and frogs.
      (2) Avoid strange fish that have a smooth slimy skin or are odd-shaped. Also avoid poisonous insects.
   f. Preparation.
      (1) Skin—clean.
      (2) Remove glands.
      (3) Check vital organs for discoloration.

(4) Boil or roast thoroughly (worms and flukes will be destroyed).

g. As all bugs, mammals, and most fish are edible, it is not necessary to recognize specific ones, except fish; but it is important to know their general and, where possible, their specific habits to obtain them for food. A few general principles concerning birds and mammals will prove helpful in trapping or hunting them.

(1) Land mammals make conspicuous signs, such as tracks, feces, runways, dens, and feeding marks that serve as indicators of their presence and relative abundance. These signs will tell whether it is worthwhile to stop or continue to a more favorable place.

(2) Birds and mammals are creatures of habit. Their normal eating habits are regular and continuously repeated. If you observe them, you can anticipate their movements. They can be trapped or hunted most successfully during their periods of activity.

(3) Birds and mammals tend to congregate in the most favorable habitats. Some of the places to look for them are:

(*a*) The edge of the woods or jungle.

(*b*) Trails, glades, and openings in forests or jungles.

(*c*) Stream and riverbanks.

(*d*) Lake and ocean shores.

(4) Birds and mammals are most active early in the morning and late in the evenings; they are generally quiet during the middle of the day.

(*a*) Hoofed animals forage both day and night.

(*b*) Many rodents and carnivora are active only at night.

Note. **Move to next station.**

E. *Fifth teaching point—hunting, fishing, and trapping.* Any trap to be effective must be constructed and set with a knowledge of animal habits. There is no "catchall" among traps. A trap set at random to catch whatever chances to come along is worthless. Decide upon the kind of animal you wish to trap, bait your snares with the kind of food it eats, and keep surroundings as natural as possible.

1. The fundamental principle of successful trapping is to determine which animal you wish to trap, what he is going to do, and then catch him doing it. It is easier to determine this for some animals than for others. Remember that wherever birds or animals are naturally abundant, trappings will prove effective.

2. Trapping hints.

a. There is no better way to attract land animals to a trap or a hide than by placing salt along a trail or waterhole.

b. A noose fastened to the end of a long pole can be used to snare an animal as it comes out of its burrow. If there is more than one entrance to the burrow, block all but one. Roosting and nesting birds can also be caught in this manner.

c. Mammals that live in hollow trees can be extracted by inserting and twisting a short forked stick. Pin the animal against the side or bottom of the hollow and then twist back. The fur and loose skin will twist the fork and the animal can be pulled out. Keep tension on the stick when withdrawing. A short fork takes a secure hold, a long one does not. These same animals can be smoked or drowned out of dens and clubbed as they emerge.

d. When all else fails, resort to fire. Game, nesting birds, and lower animal forms can be driven out of their habitats by setting fire to open grasslands. This cruel and wasteful method is not to be considered unless your life hangs in the balance.

e. Learn a few trapping techniques. If you are resourceful and if you observe the habits of wildlife, you should be able to obtain enough wild meat to sustain you. In the wilderness, resourcefulness and observation are your greatest tools.

f. Instructor now shows the class various types of snares and deadfalls that can be easily constructed.

g. If you must resort to survival tactics and have no field expedient weapons for survival, field expedients for survival can be constructed from available material and used with success. Instructor shows following weapons:

(1) Slingshot.
(2) Bow and arrow.
(3) Club.
(4) Forked stick.
(5) Fish spear.

Note. Move to next station.

F. *Sixth teaching point—shelter.* When lost or stranded, decide what is needed for safety and comfort, then look for these things. Avoid conditions that are most likely to prevent a good night's sleep. In a strange country, begin to look for a campsite two hours prior to sunset. Don't wait until dark. Consider these factors in selecting your camp:

1. Available food.
2. Good drinking water.
3. Enough level ground for your bed.
4. Protection from wind and storm.
5. Bedding and shelter material.
6. Protection from floods, wild animals, rock falls, high tides, wind and cold.
7. Concealment from enemies.
8. Absence of insect pests.
9. Firewood.

a. There are various types of shelters one can use based on available material and terrain.

(1) Natural shelters and windbreakers. Make a camp with the least expenditure of time and energy. When you find a site, examine it well for it may contain poisonous snakes, ticks, mites, scorpions, or stinging ants.

(2) Brush shelters. With a little time and effort, a brush shelter can be made of two poles lying against a log and covered with boughs or palm fronds.

(3) Snow shelters. In a cold climate, the primary purpose of a shelter is to break the air movement and retain the heat from your fire or body. The shelter should be small, windproof, and as nearly closed as possible.

(4) In desert country, you are concerned with protection from sun and heat. Wind is an important factor and cold often becomes disagreeable at night. Natural shelters, such as vegetation, overhanging rocks, and depressions afford shade provided you shift with the sun. A cover or covered trench is practical where the sand or soil is loose.

(5) Beds. A good bed serves two functions. It allows the body to relax completely and it insulates against ground chill. To do this, the bed must be dry, smooth, soft, and free of insects. Grass, sedge, dry leaves, or boughs are all good bedding material. Balsam, spruce, or hemlocks make the best bedding in cold climates.

(6) Instructor shows Soldiers various shelters that can be constructed.

Note. Move to next station.

G. *Seventh teaching point—fire.* Fire will lengthen your survival time; it enables you to keep warm, cook your food, and destroy harmful germs commonly found in food and water. You should be able to build a fire with matches under all weather conditions. No one, who may have to shift for himself in a remote area, should ever be without matches carried in a waterproof case. You can start fires by using sun and glass, bow and drill, fire thong, and fire saw. These methods are always a last resort to be used when matches are not available.

1. Remember and practice a few basic principles of fire building; you can always make a fire.
   - *a.* Select a dry sheltered area.
   - *b.* Use only the driest of tinder to start the fire.
   - *c.* Have a good supply of kindling on hand before you strike the match or use one of the methods without matches.
   - *d.* Start with a tiny fire and add fuel as the flame grows.
   - *e.* Fire needs air. Add fuel sparingly.
   - *f.* Blow lightly on burning wood.
   - *g.* Fire climbs, place fresh kindling above the flame.
   - *h.* Use dry deadwood.
2. Use judgment in the selection of a fire site.
   - *a.* Don't select a windy spot.
   - *b.* Don't build on damp ground if dry ground is available.
   - *c.* Pick a spot where the fire won't spread.
   - *d.* In rainy weather, build under a leaning tree or rock shelf.
   - *e.* When building a fire in snow, build the fire on a platform of logs, or metal; however, you can build a fire on base snow or ice.
3. All woods do not burn alike. Some scarcely burn at all; others burn quickly and make a hot flame. Some burn slowly and make good coals; some smoke, others don't. Use whatever is at hand, but where there is a choice, select the best fuels for the purpose. In general, hardwoods make a slow-burning fire with lasting coals, and soft woods make a quick, hot fire with coals that are soon spent.
4. Fires for warmth. A small fire is better than a large one for most purposes. A very small fire will warm you thoroughly if you sit or kneel over it, draping your fatigue jacket or blanket to direct all the heat upward. A reflector fire will keep you warm while you are sleeping. The base of a tree and large rocks are ready-made reflectors. Lie or sit between the fire and reflector as this will prevent you from "baking" on one side and "freezing" on the other. A reflector can be constructed of logs, snow, boughs, or sod.
5. Cooking fires.
   - *a.* When fuel is scarce, make a "hobo" stove if an empty tin can is available. Such a stove will conserve heat and fuel and is particularly serviceable in the Arctic.

b. The crisscross fire is the best all-round cooking fire because it burns down to a uniform bed of coals in a short time. The simplest fireplace consists of two rocks, two logs, or a narrow trench on which a vessel can rest over the fire below. Arrange the fireplace so that it will have a draft. If the fire does not draw well, elevate one edge of the log or stone.

6. Instructor shows various types of fires to Soldiers. Also shows field expedients used to start fires when no matches are available.

VI. Conclusion.

A—Retain attention. Instructor cites examples of how people have survived for long periods of time, applying some of the basic principles taught here.

B—Summary. Instructor summarizes teaching points, stressing "will to survive."

C—Application. This is just an introduction to survival techniques. More can be learned through your own personal experiences, Army courses, and Ranger training which will further develop survival techniques. In essence, once you have found yourself in a predicament where your knowledge and wits must help you to survive, you can be the picture of this—

Note. Demonstrator staggering by stands, ripped fatigues, one boot, bloody bandage.

or you can be this—

Note. Demonstrator sitting before a fire, cooking food in relative comfort while reading a survival manual.

APPENDIX VII

PATROL TIPS

The following patrol tips fall under three general headings: the preparation phase, which includes planning; the execution phase, and miscellaneous reminders.

a. Preparation.

(1) Make a detailed map study; know the terrain in your objective area; know your route from memory, including terrain features which will aid in navigation. Confirm these terrain features as you pass over or near them.

(2) Consider the use of difficult terrain in planning your route. You are less likely to encounter the enemy. Impassable terrain is very rare.

(3) In mountainous terrain, plan to use ridge lines for movement whenever possible, but do not plan to move along ridge tops. Stay off the skyline.

(4) Plan an offset in your route when applicable. An "offset" is planned magnetic deviation to the right or left of the straight line azimuth to an objective. Use it to verify your location right or left of the objective (fig. 61). Each degree you offset will move you about 17 meters to the right or left for each 1,000 meters you travel.

(5) When your patrol is to infiltrate enemy lines, select a rendezvous point behind enemy lines. Select an alternate rendezvous point for use if the first point is occupied by the enemy.

(6) Consider all types of grenades—fragmentation, white phosphorus, concussion, smoke, and thermite, together with the use of grenade launchers.

(7) Light automatic weapons are good on combat patrols where terrain or conditions of visibility will not permit effective employment of machineguns.

(8) Reconnaissance patrols should carry at least one automatic weapon. It provides valuable sustained firepower.

(9) Avoid taking weapons requiring different types of ammunition. It makes ammunition redistribution difficult.

(10) Clean, check, and test-fire all weapons before departure.

(11) Consider terrain vegetation. Gloves may be necessary to protect hands from briars and scratches.

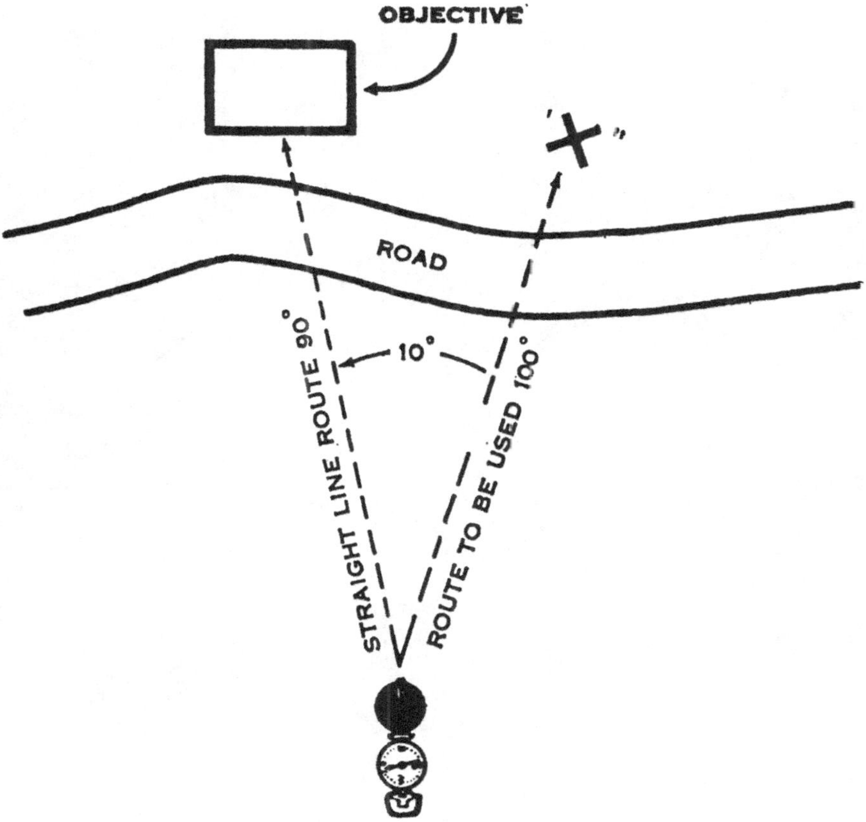

Figure 61. Use of offset. Remember the distance from "X" to the objective will vary directly with the distance to be traveled and the number of degrees offset.

(12) Consider carrying two pairs of binoculars, wire cutters, fuze crimpers, and other small items.

(13) Carry at least two flashlights for night operations.

(14) Carry extra flashlight and radio batteries on long patrols.

(15) Every man should carry his canteen and poncho. Consider having each man carry two canteens on long patrols. If special circumstances make it undesirable for every man to carry his canteen and poncho, carry at least two or more in the patrol.

(16) Ponchos can be used to make litters, construct rafts, conceal lights, and as shelters (fig. 62).

(17) Have every man carry an extra pair of socks.

(18) Carry a sharp knife on the harness or concealed in a boot. The Boy Scout type knife is good to carry in the pocket.

(19) Harness should be worn when the belt is worn.

(20) Carry individual weapons cleaning equipment on all patrols. Check to see that the oiler is full and that patches are carried.

Figure 62. Where possible, insure that all patrol members receive a thorough map orientation.

(21) Consider the use of scout dogs if they are available.

(22) A length of rope, secured to the harness, can be used for binding prisoners, climbing or descending obstacles, and crossing streams.

(23) Two pieces of luminous tape, each about the size of a lieutenant's bar, worn on the back of the collar, aid in control and movement on dark nights. Turn the collar down when close to the enemy. The tape can also be worn on the back of the cap, but cover or remove it when close to the enemy.

(24) Use friction tape to secure rifle swivels, sling, and other items which might rattle.

(25) Be sure to camouflage the back of your neck, behind your ears, and the back of your hands.

(26) A clear acetate sheet placed over luminous tape can be used to make roughstrip maps at night. The map will glow in the dark, making the use of lights unnecessary. Use a grease pencil so information can be easily erased.

(27) Machinegun ammunition, minus the box, can be carried conveniently in its package container. It can be fed into the gun from the container.

(28) Provide for security by assigning every man an area of responsibility.

(29) Designate at least two pacers and use the average of their individual counts.

(30) Fold and prepare maps before departing to facilitate map checks en route.

(31) Preset compasses before departing.

(32) Prepare a list of coordination questions to be asked at the position from which you depart.

(33) When appropriate, arrange to have a light aircraft reconnoiter ahead of your patrol to keep you informed of any enemy activity or ambushes along your route.

(34) Take your assistant patrol leader or element leaders with you on reconnaissance.

(35) Prearrange and rehearse all signals to be used. Keep signals simple.

(36) Plan time for your patrol members to dark-adapt their eyes, if you have a night patrol.

(37) Use available visual aids in issuing your patrol leader's order. The use of a blanket board, blackboard, or a sketch on the ground is helpful.

(38) Do not carry maps marked with information that might aid the enemy.

(39) Conduct rehearsals on terrain similar to that over which you will operate. Conduct day and night rehearsals for a night patrol.

(40) Inspect your patrol carefully before rehearsals and before departure. Question men to check their knowledge and understanding of the actions planned.

b. Execution.

(1) Have your assistant patrol leader check and count the patrol through friendly positions.

(2) On small patrols, the count should be sent up automatically after each halt or passage of a danger area. In large patrols, use the chain of command to account for men.

(3) Use the point man as a point and not as a compass man; he is primarily concerned with security. Have the second or third man responsible for navigation. Check navigation frequently. On long patrols, change both point and compass men occasionally.

(4) Use a code word or a password forward of friendly positions. (Other than the assigned challenge and password.)

(5) At halts and during movement, odd numbered men observe to the left, even numbered men to the right.
(6) Weapons are always carried at a ready position.
(7) Cut enemy wire only when necessary. Make a reconnaissance first.
(8) When moving at night take advantage of any noises such as wind, vehicles, planes, shelling, battle sounds, and even sounds caused by insects.
(9) Stay off roads and trails for movement unless their use is deemed absolutely necessary.
(10) Use stars to aid in navigation. Facilitate movement during daylight hours, especially in dense terrain, by using night compass settings.
(11) When in close proximity to the enemy main battle position, avoid lateral movement across its front.

Figure 63. "When we march, we keep moving till dark, so as to give the enemy the least possible chance at us."

(12) When men have difficulty staying awake on security and at halts, minimize the number of halts and have the men assume a kneeling rather than prone position.
(13) Over short distances, such as the width of a road, the compass can be used for signaling at night. A piece of luminous tape can also be used for this purpose.
(14) There are several acceptable methods of crossing roads. Whatever the method used, the basic principles of reconnaissance and security apply. Some of the accepted methods are—

(a) Patrol can form a skirmish line and advance quickly across the road.

(b) The entire patrol can form a file, following the footsteps of the man in front in order to minimize detection of footprints.

(c) Men cross the road a few at a time until patrol is across.

Crossing roads in enemy territory is merely a matter of common sense. Each situation may dictate a different method. You will not violate the established procedures providing you apply the proper reconnaissance prior to crossing the road; establish adequate security and move silently and quickly to avoid detection. A main point of consideration in any road crossing is control of your unit.

(15) Crossing streams is similar to crossing roads; reconnaissance and security are both necessary.

(16) If it is necessary to leave a wounded man to be picked up on your return trip, leave another man with him, if possible. Walking wounded return on their own to friendly lines, if feasible. When close to the enemy, remove the wounded from the immediate area before applying first aid.

(17) Avoid all human habitations.

c. *Miscellaneous.*

(1) Keep the cutting edge of the entrenching tool extremely sharp. It is a good silent weapon and can be used in lieu of a machete.

(2) A razor blade or sharp knife and a piece of cord are good substitutes for a snakebite kit.

(3) A garrote can be used for killing a sentry or capturing a prisoner.

(4) Binoculars increase visibility at night.

(5) Do not jeopardize security by letting earflaps and hoods interfere with the hearing ability of the patrol.

(6) When on patrol, pass on simple instructions, allow time for dissemination, then execute.

(7) Keep talking to a minimum. Use arm-and-hand signals to the maximum.

(8) When reconnoitering enemy positions, keep the covering force within supporting distance of the reconnaissance element.

(9) Never throw trash on ground while on patrol. Bury and camouflage it to prevent detection by the enemy.

(10) When possible, allow men to sleep on long patrols, but maintain proper security.

(11) When contacting friendly agents, such as partisans, never take the entire patrol to make contact with them. Have one man make the contact and cover him.

(12) Do not let the desire for personal comforts endanger the patrol and the accomplishment of the mission.

(13) The best nights for patrols are dark, rainy, and windy nights.

Figure 64. "Let the enemy come till he is close enough to touch. Then let him have it, and jump out and finish him up with your hatchet." MAJOR ROBERT ROGERS—1756.

APPENDIX VIII

GENERAL CONSIDERATIONS OF PATROLLING

1. General

a. Units can establish standing operating procedures for patrolling to insure uniform procedures for patrolling operations, promote more efficient patrolling, expedite administration and coordination in support of the unit patrolling effort.

b. Unit patrolling is of great importance to the combat intelligence effort and the success of pending operations. The patrol efforts will be carefully integrated with the collection plan of the unit. Rifle companies initiate and report ground reconnaissance as required without orders from unit headquarters. Reconnaissance patrolling is coordinated by S2. Subunits submit patrol plans to the S2 daily or as initiated.

c. Patrols will be of two types. Size will vary with the mission.

(1) *Reconnaissance.* Reconnaissance patrols move to specified points or areas, gather required information through observation, and report information obtained. They will avoid enemy contact whenever possible, fighting only when necessary to accomplish the mission. A five- to eight-man patrol is considered a normal size reconnaissance patrol.

(2) *Combat.* Combat patrols are heavily armed detachments sent out to kill or capture the enemy, destroy his equipment or materiel, or installations. The size of a combat patrol is consistent with the assigned mission.

d. In assigning patrol missions, the following factors will be considered: location, disposition, current and projected tactical missions of the units; the state of training, morale, fatigue, combat effectiveness; and recent battle history. When assigning missions which are usually difficult or hazardous, it is particularly important to consider carefully all the factors involved.

e. To insure successful preparation and conduct of patrols, emphasis must be placed on individual tactical training. Such subjects as map reading, battlefield movement, observation, camouflage, maintenance of direction, use of supporting fires, and the use of specialized equipment such as infrared devices must be integrated into tactical training exercises wherever possible. At least ⅓ of the training of this type will be conducted at night. Small unit patrolling exercises will be conducted frequently for squad and platoon.

2. Planning and Preparing Patrols

a. The S2 will plan missions for reconnaissance patrols and the S3 for the combat patrols. Both officers will work in close liaison to insure a coordinated effort.

b. The Daily Patrol Plan (Annex A) will include all patrols carried out in the unit area, including patrols initiated by the subordinate units. This plan, compiled daily by the S2 and approved by the commander, will be sent to higher, lower, adjacent, and supporting elements for information and coordination.

c. Patrol personnel will be notified of patrol missions in sufficient time to permit the accomplishment of daylight reconnaissance and other necessary preparation prior to departure. Patrols, when possible, will be provided with latest maps, aerial photos and sketches showing routes, and objective areas. Coordination with the patrol leader will be made for ground reconnaissance and, whenever possible, air reconnaissance.

d. Patrol order normally will be issued at the CP and/or suitable observation point. The S2 or S3, after having coordinated with the staff and support agencies, will issue the order to patrol leaders and at least one other member of the patrol. The order should include information contained in Annex B. The S2 will normally issue the order to reconnaissance patrols. Combat patrols will receive their order from the S3; however, prior to their departure they will also be briefed by the S2 on the terrain and enemy situation.

e. Patrols may be augmented with special equipment such as infrared vision devices, sniperscopes, scout dogs, special weapons and special transportation such as helicopters, according to dictates of weather, terrain, and missions assigned. Armored vests will be worn on short-range combat patrols. Amount of ammunition carried is dependent on the mission, distance to be covered, terrain, and weather. A minimum of two fragmentation grenades per individual will be carried. The use of helicopters must be rehearsed and coordinated as to time, rendezvous areas, withdrawal landing site, loading plans, and landing lights. The use of scout dogs will be coordinated with the dog platoon commander. The responsible staff officer will insure this coordination is effected.

f. The individual Soldier's preparation for a patrol will include but not be limited to: camouflage of his equipment; camouflage of exposed body surfaces; and a check for serviceability of equipment. Check for serviceability of equipment includes test firing of weapons and tying down equipment which might rattle. At night, sufficient time will be provided prior to departure for patrol members to accustom their eyes to darkness. All personal effects which would aid the enemy in identifying units will be removed. Only identification tags and Geneva Convention Identification Cards will be carried.

g. The responsible commanders and/or staff officers will supervise rehearsals of actions at the objective. Mockup areas of the objective will be arranged. Communications will be checked during these rehearsals. Prior to the departure, the patrol leader and his unit commander or responsible staff officer, will inspect the patrol.

h. The S2 and S3 coordinate patrol movements with patrols and operations of higher and adjacent units. Patrols normally will be controlled by next higher headquarters. Control will be exercised by assignment of general routes, checkpoints, and times to report. Radio will be the primary means of communication, but wire may be used. Emergency signals, such as pyrotechnics, will be prearranged. Deviation from a prescribed route must be reported as soon as possible.

i. Fire support plans for each patrol will be coordinated to eliminate duplication and insure safety.

j. The patrol leader will contact the responsible staff officer upon completion of his plan to coordinate all details and arrange for any additional support desired.

k. The patrol leader's order to the patrol must be clear, concise, and in sufficient detail to insure that each member understands his particular role in the patrol operation.

3. Conduct of Patrols

(FM 7-15)

a. Patrols will physically contact friendly troops occupying the position closest to the point of departure and inform the company commander of the general route, time of return, number of men in the patrol, and special signals, if any. In the event the patrol is overdue or communications fail, all outguards will be notified by the responsible staff officer or unit commander.

b. Patrols will avoid all enemy contact and will fight only to protect themselves or to accomplish their missions.

c. The assignment of one mission does not preclude the collection of incidental information during the patrol. All members must be alert for any information which may be of value to the unit. This should be pointed out in the patrol order.

d. Patrols will attempt to take prisoners only if such action can be taken without jeopardizing the mission. Prisoners may be bound, gagged, and taken along with the patrol; or, they may be bound, gagged, and hidden to be brought in on the return trip. If a prisoner must be left behind, his exact location will be reported to unit headquarters. All prisoners will be disarmed, searched, and relieved of all documents. When possible, unit identification will be ascertained.

623989°—62——19

4. After Action

a. All patrols will be debriefed. The S2 is responsible for debriefing patrols immediately upon return to the CP area. Normally, patrol leaders and members of the patrol will be allowed to give an oral report. The S2 or a member of the intelligence section will prepare the patrol report form.

b. The S2 or staff member on duty will report to the commander the return of each patrol and a synopsis of the results of the mission. A complete patrol report will be furnished higher headquarters. Maps and overlays in explanation will accompany this report.

c. All pertinent information reported by patrols will be disseminated to the subordinate units by the fastest means available consistent with security.

d. Every attempt will be made to provide hot meals, facilities for washing, medical attention, and rest in a comfortable area for all members of a returning patrol. The morale and general welfare will be adequately provided for by proper planning prior to the patrol's return. The CO, Headquarters and Headquarters Company, should be responsible for rest and rehabilitation facilities.

ANNEXES: A—Daily Patrol Plan
B—Patrol Order Format

ANNEX A

(Daily Patrol Plan) TO GENERAL CONSIDERATIONS OF PATROLLING

The following format will be used for daily patrol plans:

PATROL PLAN

DATE________________

	No 1	No 2	No 3	No 4
PATROL NUMBER				
UNIT ASSIGNED				
MISSION				
TYPE				
SIZE				
ROUTE OR ZONE OVERLAY				
TIME OF DEPARTURE				
TIME OF RETURN				
LOCATION OF CHECKPOINTS				
METHOD OF REPORTING				
FIRE SUPPORT PLAN				
SPECIAL EQUIPMENT				

ANNEX B

(Patrol Order Format) TO GENERAL CONSIDERATIONS OF PATROLLING

1. SITUATION
   a. Enemy Information.
      (1) Known or suspected enemy positions.
      (2) Enemy patrol activities.
      (3) Known or suspected enemy ambush sites.
      (4) Terrain and weather and their probable effects on the patrol.
   b. Friendly Information.
      (1) Mission of the unit to include any planned operations of organic, attached, or supporting units which might affect the patrol.
      (2) Missions and routes of other friendly patrols operating in the vicinity.
      (3) Location and planned actions of adjacent units.
   c. Attachments.
      (1) Specialist personnel.
      (2) Forward observers.
2. MISSION
   (A brief, concise, and complete statement of what the patrol is to accomplish.)
3. EXECUTION
   a. Concept of the Operation.
      (1) Recommended size of the patrol.
      (2) General route (usually indicated by the designation of checkpoints.)
      (3) Fire support available.
   b. Coordinating Instructions.
      (1) Time of departure and return.
      (2) Departure from and re-entry of friendly area (s).
      (3) Detailed fire support plan to support the patrol.
      (4) Time and place for debriefing.
4. ADMINISTRATION AND LOGISTICS
   a. Special Equipment Available for Patrol.
   b. Transportation.
   c. Instructions for Handling of Wounded and Prisoners.
   d. Rehearsal Areas.
5. COMMAND AND SIGNAL
   a. Signal.
      (1) Call signs and frequencies.
      (2) Special instructions for use of communications.
      (3) Reports to be made and methods of transmission.

(4) Challenge and password.

b. Command. The location and definite designations of the agency and the individuals of next higher headquarters who will provide information and support to the patrol during the preparation and execution phases.

Note. **When this format is used as an annex to an SOP for patrolling or as a patrol order, it will be in complete accordance with FM 7-40.**

APPENDIX IX

EXAMPLE PROBLEM OUTLINE TO A TRAINING MEMORANDUM

A problem outline or scenario, similar to the one below (problem #3), normally will be included in a training memorandum when the Ranger program is initiated by a higher headquarters. This type of problem outline greatly assists the principal instructor and assures that the commander's concept of instruction is effectively executed.

ANNEX __________ TO BATTLE GROUP TRAINING MEMORANDUM ____________________

Ranger Field Exercises

AMBUSH PATROL

1. GENERAL SITUATION

a. Maps of problem area, 1:25,000, Incl. 1.

b. Your company is presently in division reserve and has received training in Ranger type operations. The 10th Division has been opposing the aggressor in northern Georgia along the mountainous Tennessee Valley Divide Road. The latest division intelligence summary follows:

Aggressor continues to defend strong points along TVD. Patrolling between strong points stepped up during the past 48 hours. Motor convoys on all main supply routes now operating around-the-clock, continuing to bring up both personnel and equipment. Size and composition from two to four 2½-ton type vehicles preceded by scout ¼-ton closed column. Rate of movement slow due to condition of mountain roads. Scout ¼-ton precedes convoy from 1 to 3 minutes. Combat patrols from 2d Ranger Bn raided and destroyed __________ Bn, seized CP installations at _________________, _________________, and _________________. Prisoner raids conducted by division patrols gained information of impending convoys. Size and composition of raided CP installations confirmed previous reports of enemy positions containing 2 and 4 log and sandbag bunkers, approximately 2.5 x 3 meters, well dug in, and 10 to 20 men on position. Resistance at all CP locations was fanatical. Collar insignia indicates no new units are being sent to the front. Weather continues ________________

No indication of withdrawal. Aggressor capable of defending or attacking from present positions.

2. INITIAL SITUATION

The Soldiers are gathered in the briefing area and the principal instructor conducts the briefing using maps, charts, blowups, and other training aids. The briefing is as follows: (PI conducts briefings.)

FIRST REQUIREMENT. Actions and orders of leaders.

SOLUTION. Patrol leaders (each Soldier initially) prepare warning order, form tentative plan, issue warning order, make reconnaissance, coordinate and issue patrol order.

3. SECOND SITUATION

The patrol crosses the LD, runs into harassing artillery fire in "no man's land," receives first casualty.

SECOND REQUIREMENT. Final coordination and passage of friendly lines. Actions and orders of leaders, reaction of patrol to artillery and mortar fire, and formations.

SOLUTION. Patrol leader coordinates passage with FFL commander, last outpost, etc. Counts men through lines, minefields, wire, and traps; moves patrol rapidly through harassing fire. Make disposition of casualty.

4. THIRD SITUATION

The casualty is a minor wound and is able to continue. The patrol approaches the enemy positions. If the patrol is obviously noisy and is detected it receives enemy fire, both automatic weapons and mortar. If the patrol used sufficient stealth, it may breach the positions of the enemy without detection.

THIRD REQUIREMENT. Passage of enemy lines, reactions of the patrol to enemy fire, actions and orders of all leaders, reorganization of the patrol behind enemy lines.

SOLUTION. It is not likely that automatic weapons or mortar fire will pin down the patrol and therefore it should be able to move around the strong points and pass through the lines. Proper commands and signals for rallying after the firing should be employed. Use of reconnaissance, wire cutters, and stealth aid patrol movements through the lines.

5. FOURTH SITUATION

The patrol organizes in rear of the enemy positions and continues its movement to the clandestine bivouac area.

FOURTH REQUIREMENT. Use of pacers, point men, and

night patrolling methods employed. Actions and orders of leaders, actions of patrol members, conduct of the patrol.

SOLUTION. If patrol leader has been changed or declared a casualty, new patrol leader may change pacers, point men, men with heavy loads, etc. This is done far to the rear of the last known enemy position and after a reconnaissance. Night navigation using compass, stars, and guiding on prominent terrain features aids the patrol. Pacers are checked. Counts are passed from the rear. Patrol members are alert, move rapidly but silently.

6. FIFTH SITUATION

Patrol moves along planned route guiding on terrain features, stars, etc. Patrol arrives vicinity of clandestine bivouac area.

FIFTH REQUIREMENT. Actions and orders of the patrol leader, actions of the patrol, conduct of the bivouac.

SOLUTION. See problem number 3 checklist. Reports are made by radio.

7. SIXTH SITUATION

At dawn patrol completes rest period and insures that all members are alert. The ambush site is put under observation, security is placed further out from the bivouac site. Detailed reconnaissance of the ambush site is made. Plans are reaffirmed and visually confirmed by study of the terrain. Final confirmation of each element of the patrol is made and preparations to move to the ambush site are completed. Prior to complete darkness, the patrol deploys to the ambush site.

SIXTH REQUIREMENT. Actions and orders of leaders, actions of patrol members, conduct of the movement, and deployment of position.

SOLUTION. Patrol leader physically shows elements of the patrol where they will be located for the ambush by personally pointing out areas. Patrol leader reviews communication plan. AN/PRC-6 radio nets are opened to control movement into position and make final coordination. Visual signals are reviewed; pyrotechnics, and crimped cartridges are issued to key personnel. Patrol crosses road after reconnaissance and camouflage or removes evidence of presence. Security elements are employed at advantageous positions. Roving patrol is placed in rear for security. Assault element is placed in position. Patrol leader is with the assault element. He checks ammunition, weapons, and equipment and supervises patrol to insure alertness and silence.

8. SEVENTH SITUATION

Two enemy vehicles approach the ambush site. Later a five-vehicle convoy is observed. The convoy consists of a $\frac{1}{4}$-ton vehicle (scout) followed by four $2\frac{1}{2}$-ton type trucks. Two trucks carry equipment and troops and two trucks are marked as explosives (in aggressor language).

SEVENTH REQUIREMENT. Actions of security elements, actions and orders of leaders, actions of patrol.

SOLUTION. The security element reports approach of two vehicles. Patrol leader orders by phone and radio the two vehicles be allowed to pass. Observation of the vehicles is noted for the report later. The larger convoy is marked for destruction. The scout vehicle is allowed to proceed through the ambush. The patrol leader orders that all personnel will be killed from position and that fire will not be directed into ammunition vehicles unless they catch fire. The ambush is conducted vigorously. The search party insures all personnel are dead in the area, places thermite grenades on the truck engines, and other demolitions throughout the convoy to insure destruction of the ammunition trucks after the patrol is away from the ambush. The explosion of the vehicles is observed from the rally point. The scout car which returned initially while the search party was at work was destroyed by the security. The patrol commences movements to the contact position. Radio contact is made to report "CONDITION BLACK."

9. EIGHTH SITUATION

The ambush successfully completed, all elements successfully withdraw to the rally point in rear of the ambush site. The patrol is reorganized for the movement to the rendezvous area.

EIGHTH REQUIREMENT. Actions and orders of patrol leader, conduct of the patrol en route to the rendezvous area, conduct of the patrol at the rendezvous area, contact with the friendly patrol.

SOLUTION. Casualties, if any, are assessed and disposed of accordingly. Organization for movement should be basically similar to previous movement, except weapons ammunition, equipment loads, etc., may be shifted, redistributed, or exchanged. Pacers and point men are employed. If arrival in the rendezvous area is prior to the friendly patrol, the area is reconnoitered and a defense perimeter employed. If upon approach to the area, a signal is observed to indicate the friendly patrol, precautions are taken for proper identification before making contact. This is done to avoid ambush by enemy. Making proper use of the challenge and password, contact is made and the patrol is returned to the debriefing point.

APPENDIX X

PROBLEM COMMUNICATIONS SUPPORT

INTRODUCTION

Communications training is a continuous process and is an integral part of the Ranger program. Radio and wire nets are established primarily to support the play of the problem and secondly for con-

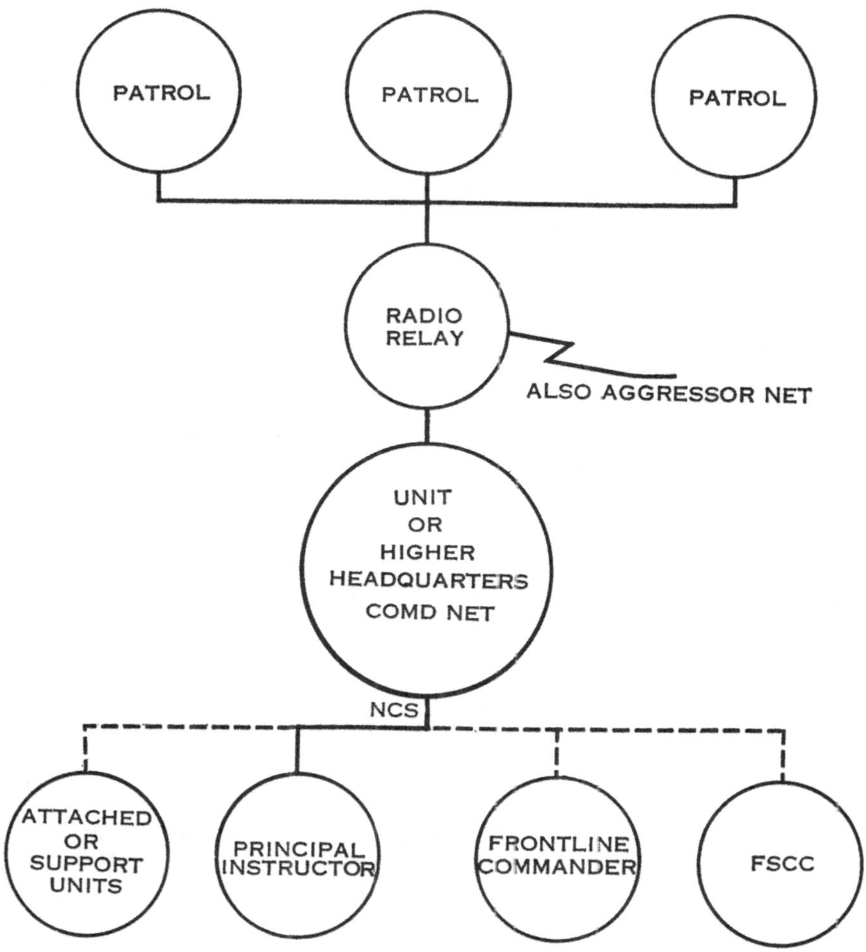

1 Patrol problem radio net

Figure 65. Radio nets.

trol and safety reasons. The principal instructor will encounter two main problem areas in the effective ranges of standard FM line of sight radios: constant rotation of radio man positions require that all Soldiers be qualified operators; they must understand the limitations, maintenance, and operational procedure of radios used in the patrols. Communications training and relay stations can offset these difficulties to a great degree.

The radio nets (fig. 65) are generally applicable to patrolling problems. Wire nets (not shown) can be used to effect communications

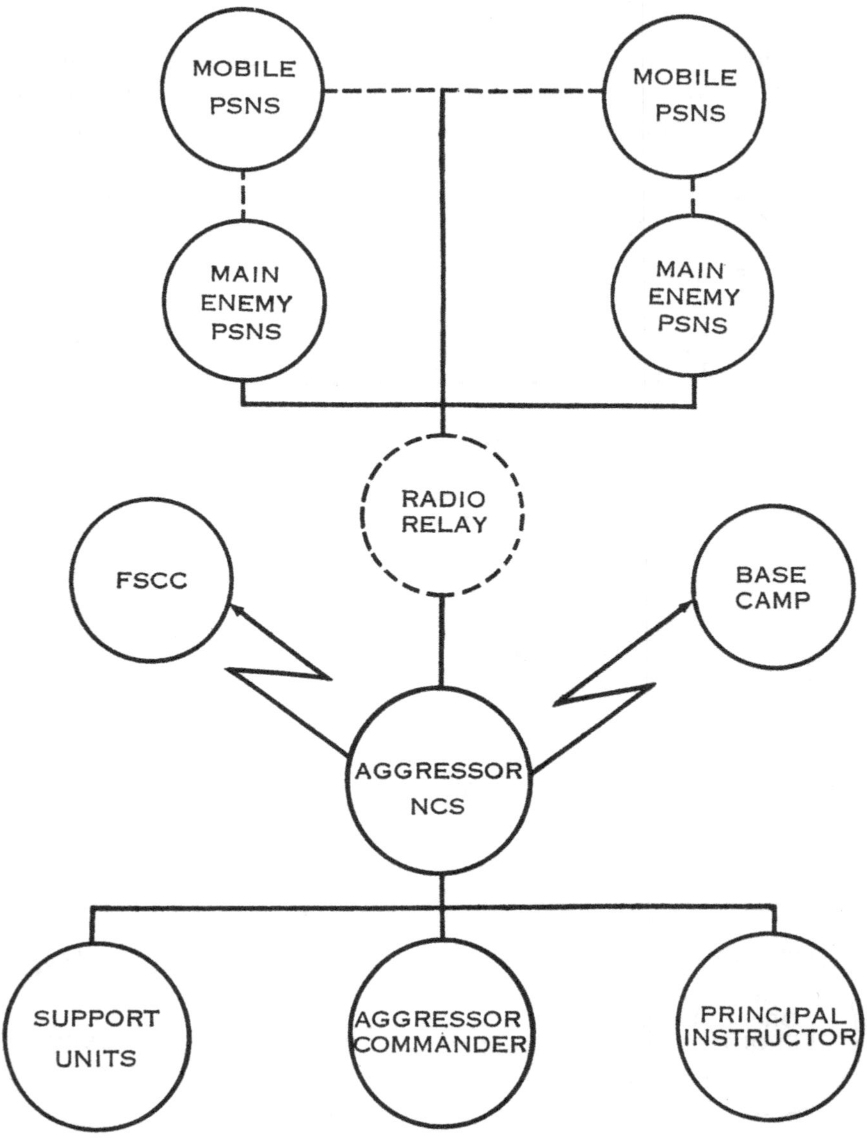

2 Type aggressor radio net

Figure 65—Continued.

with the main aggressor positions during tactical problems that require dependable semipermanent means of communications.

Note 1. Each patrol will normally carry one or two AN/PRC–10 radios for communication with higher headquarters and/or the FSCC. One of these radios can be carried as a spare and used in lieu of the AN/PRC–6 on ambush or similar actions for interpatrol communication.

2. Higher headquarters will normally be the base camp.

3. Radio relay may also support aggressor and fire control nets when applicable.

4. Attachments may include pathfinder support, aircraft, etc.

5. The aggressor net may also be used as the problem administrative net.

6. Mobile aggressor positions are normally the enemy FEBA strong points, patrols, convoys, etc.

7. The FSCC uses the aggressor positions to fire support missions.

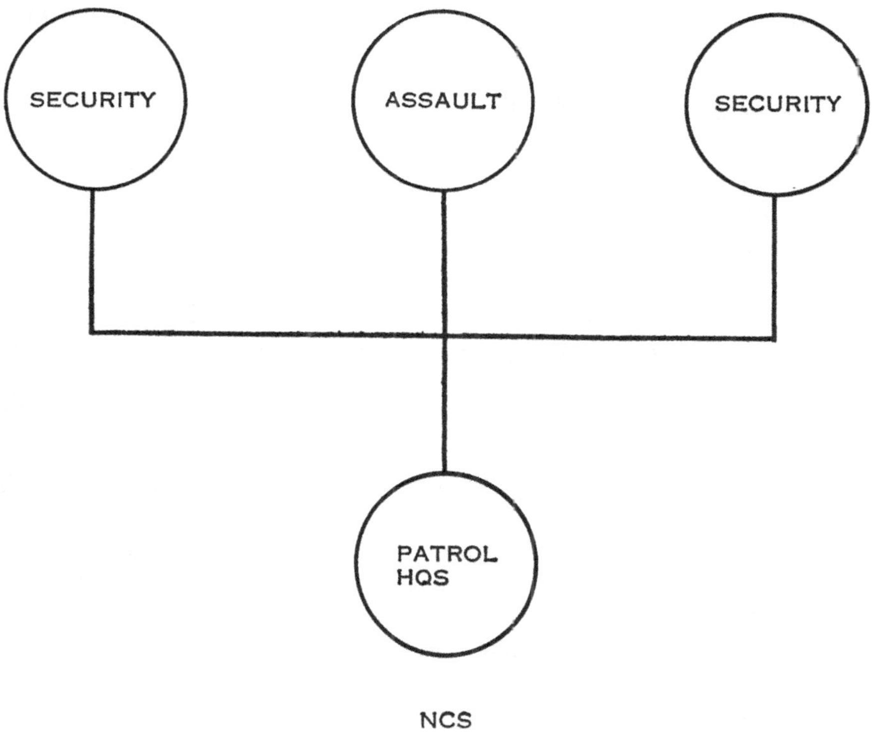

8 Type interpatrol radio net

Figure 65—Continued.

APPENDIX XI

GENERAL CRITIQUE CHECKLIST

1. Introduction

a. The following general critique points are designed to assist the lane instructor in performing his duties.

b. A special critique checklist should be prepared to suit the needs of each problem and may include—

(1) A recommended equipment list based on the mission to be conducted.

(2) Special doctrine or techniques associated with the problem.

(3) Salient points which the PI desires to emphasize before, during, and after the problem.

2. Preparation and Planning

a. Troop-leading steps will be followed by the patrol leader.

b. The warning order will include all points listed in FM 7–15. In addition—

(1) The mission will state the location of objectives in terms of specific coordinates.

(2) The equipment for all the patrols should include one entrenching tool per squad-size unit—for use as a silent weapon, in sanitation, police, for burial of dead, etc.

(3) All individuals in the patrol will be armed and no one will be excused from carrying a weapon for any reason.

(4) The uniform will include "dogs tags" and ID card for each individual.

(5) Maximum use will be made of all available training aids.

c. Fire support coordination will include all items specified in paragraphs 3*d*1 to 10 and 3*e* of SOP for fire support coordination.

d. Friendly frontlines (FFL) coordination:

(1) Company commander or S3 coordination will normally include the following:

(*a*) Introduction to coordinator and size of patrol.

(*b*) A brief statement of patrol purpose, route planned, location of objective, and times of departure and return.

(*c*) The ground location of the LP to confirm coordination received in briefing.

(*d*) The wire, enemy or friendly; its location, gaps, and type;

minefields, enemy or friendly; type markings, and safe lanes.

(*e*) All recent known enemy activity and positions in the patrol's area of operation.

(*f*) The location of friendly frontlines—to avoid paralleling.

(*g*) All medical support, fire support—small arms, heavy weapons, etc.

(*h*) The use of ground for rehearsals, messing, rally points, etc.

(*i*) Use FFL radio net in emergency only—call signs, frequencies, code words, etc.

(*j*) CO's presence when patrol returns.

(*k*) Corps challenge and password, and special password(s) for patrol use forward of FFL.

(*l*) Guides from detrucking point to LP's and the routes to be used.

(2) LP coordination will normally include (coordinator will consider rank, duties of man on LP when questioning him):

(*a*) Tenure of duty and passing on information to his relief.

(*b*) The challenge and password to include special password.

(*c*) All recent enemy activity, time of departure and return, and size of patrol.

e. Aerial reconnaissance and aircraft coordination.

(1) The coordinator must know area to be reconnoitered and must brief and coordinate with the pilot prior to explanation of exercise.

(2) Individual equipment for aerial reconnaissance will include: harness—complete, compass, binoculars, weapon with ammunition, map.

f. Miscellaneous coordination will include—

(1) Coordination with other patrols in the vicinity, use of artillery, and special passwords.

(2) When applicable, coordination with adjacent units on routes, use of artillery and special passwords.

g. The patrol order will follow the order prescribed on the patrol leader's Order Card. In addition—

(1) Training aids will be used to the maximum.

(2) The mission will specifically state coordinates of objective in the warning order. Passage of FFL (paragraph 3*c*3, Patrol Order) will include plan for security halt forward of FFL.

(3) The formation (paragraph 3*c*6, Patrol Order) should not place the assistant patrol leader as last man in column for patrols, since he is concerned with control, security, etc.

(4) Actions at danger areas (paragraph 5*a*3, Patrol Order) must include special passwords planned for use forward of friendly frontlines.

(5) Patrol order will include all information obtained in coordination with the FFL, FSCC, air and adjacent unit.

Note. Additional information on planning and preparation is contained in Appendix VII, Patrol Tips.

3. Conduct of the Patrol

a. Movement.

(1) Passage of friendly positions.

(*a*) Coordinate with the CO upon arrival in the area for any changes in the enemy or friendly situation since previous coordination.

(*b*) Physically show initial rallying point to the patrol.

(*c*) Assistant patrol leader counts patrol out of point of departure.

(*d*) At night, patrol will have security halt forward of friendly positions. Patrol is absolutely quiet with all-round security. Normally, in daytime it is unnecessary to have security halt if visibility is good and the point is properly utilized.

(*e*) If tactically sound, a 5-minute break should be taken after the first 15 minutes of movement. This break may be taken prior to departing FFL if movement was strenuous.

(2) Routes (to include alternate routes which must always be planned):

(*a*) Selection.

1. Make a thorough map reconnaissance and know the number of terrain features in the area.

2. Avoid roads and trails, as a general rule.

3. Do not use roads and trails between main battle areas.

4. Unused trails may be used with caution when deep in enemy territory—beyond 3,000 to 5,000 meters rear of FEBA.

5. Do not guide on roads (if you can see the road then you can be seen from the road).

6. Do not use major ridge lines between enemy and friendly positions.

7. Pass above gaps and below key terrain features.

8. Route selected should avoid known enemy positions.

9. Avoid all built-up areas.

(*b*) Additional information on route selection is contained in FM 7–15.

b. Navigation.

(1) Use map and compass together.

(*a*) Orient map to the ground, not the ground to the map.

(*b*) Check marginal data, contour interval varies.

(2) Use two or more pace men.

(*a*) Use average pace.

(*b*) Two or three times as many paces are required in difficult terrain as on level ground.

(3) Always change map distance to ground distance.

Note. Additional information on navigation is contained in FM 7–15, appendix VII.

c. Rate of March.

(1) Conserve strength.

(2) You must reach the objective in a condition to fight in order to accomplish the mission.

(3) Use proper mountain walking techniques when applicable.

d. Security.

(1) Don't just talk about security—DO IT.

(2) Move only as fast as security and control will allow.

(3) Select capable personnel for point and compass men. Patrol leader must frequently check accuracy of this team.

(4) Speed will sometimes provide security.

(5) Use fakes and ruses when crossing FEBA.

Note. Additional information on security is contained in FM 7–15 and appendix VII.

e. Danger Areas. Include all areas listed in the critique for patrol order. Additional information is given in FM 7–15 and appendix VII.

f. Rallying Points.

(1) All rallying points are termed *tentative* rallying points until they are reached, found to be suitable, and designated.

(2) When you designate a rallying point, halt your patrol and tell them, "This is a rallying point." Point out identifying features. Be sure the information is passed to all patrol members.

(*a*) *Types.* There are three types of rallying points:

1. Initial rallying point. A point within friendly areas where the patrol can rally if it becomes scattered before departing friendly areas or before reaching the first rallying point en route. The initial rallying point must be coordinated with the commander or leader in whose area it lies.

2. Rallying points en route. Rallying points between friendly areas and your objective.

3. Objective rallying point. A rallying point near the objective where the patrol reassembles after the mission is accomplished. Where appropriate, this can be used as the point from which the leader's reconnaissance is conducted and from which elements and teams move into position to accomplish the mission.

(*b*) *Selecting rallying points.*

1. Select likely locations for rallying points during your reconnaisance or map study. Designate them as tentative rallying points in your patrol leader's order. Remember that they may prove unsuitable and must be confirmed and announced when you reach them.

2. *Always* select a tentative initial rallying point and a tentative objective rallying point. If you cannot locate suitable areas during reconnaissance or map study, designate these two tentative rallying points by grid coordinates or in relation to terrain features.

3. Select additional rallying points en route as you reach suitable locations.

4. When you reach a danger area you cannot bypass, such as a trail or stream, select a rallying point on both the near and far sides. If good locations are not available, designate the rallying points in relation to the danger area. For example, say, ". . . fifty meters this side of the trail," or, ". . . fifty meters beyond the stream."

(a) Personnel who reconnoiter the danger area must also check beyond the danger area for a suitable rallying point. On the basis of their report, you designate the rallying point.

(b) If the patrol's crossing of the danger area is interrupted, or if a portion is separated from the patrol, all members proceed to the rallying point on the far side as soon as possible.

(*c*) *Use of rallying points.*

1. The *initial rallying point* and *rallying points en route* are selected to prevent complete failure of the patrol if it is unavoidably dispersed. *YOU MUST MAKE EVERY EFFORT TO MAINTAIN CONTROL TO AVOID USING THESE RALLYING POINTS.* The success of your patrol is jeopardized if it is dispersed and forced to rally.

2. The objective rallying point, however, helps the patrol to reassemble after the various elements and teams have separated to perform their assigned missions.

3. If dispersed within the friendly area, patrol members assemble at the initial rallying point.

4. If dispersed between the friendly area and the first rallying point en route, patrol members move to the initial rallying point or to the first rallying point en route. *You must state the rallying point to be used in your patrol leader's order.* Your decision is based on careful consideration of all circumstances.

(a) Return to the initial rallying point may be extremely difficult due to mines, wire, or the enemy situation.

(b) Forward movement to the first rallying point en route may also be difficult, impractical, or impossible. The point you have selected may be mined or occupied by the enemy. The cause of dispersal, such as enemy contact, may prevent forward movement. Without maps and compasses, your men may not be able to locate the point.

5. If dispersed between rallying points en route, patrol members return to the last rallying point or move to the next tentative point. *You must make and announce this decision at each rallying point.* As before, your decision is based on careful consideration of all circumstances.

(*d*) *Actions at rallying points.* Plan the actions to be taken at rallying points and instruct your patrol accordingly.

At the initial rallying point and rallying points en route, you must provide for the *continuation of the patrol* as long as there is a reasonable chance to accomplish the mission. For example, you may plan—

1. For the patrol to wait until a specified portion of the men have arrived and then proceed with the mission under the senior man present. This plan could be used for a reconnaissance patrol where one or two men may be able to accomplish the mission.

2. For the patrol to wait for a specified period, after which the senior man present will determine actions to be taken, based on personnel and equipment present. This could be the plan when a minimum number of men or certain items of equipment, or both, are essential to accomplishment of the mission.

g. Patrol Check. The patrol leader should check his patrol, both on the move and at halts (most often on the move).

h. Control. See FM 7–15.

i. Actions at the Main Battle Area.

(1) Maximum stealth must be employed.

(2) If penetration by stealth is unsuccessful, an alternate aggressive plan must be executed. (An alternate plan must always exist and be known by patrol members.)

(3) If aggressor contact is light, patrol should move aggressively through the enemy line.

(4) Available, preplanned artillery should be used to assist in crossing the FEBA.

j. Passing Obstacles. Information contained in FM 7–15.

k. Return to Friendly Lines.

(1) Re-entering FFL: Issue spot report to commanding officer to include—

(*a*) Status of patrol, including number and names of missing personnel.

(*b*) Summary of pertinent enemy information by the patrol.

(2) Debriefing: See FM 7–15.

4. Miscellaneous

a. Was the warning order clear, complete, and concise?

b. Were adequate items of equipment and ammunition selected?

c. Did the patrol leader make a thorough map study prior to his reconnaissance?

d. Was the reconnaissance complete, including coordination with frontline personnel?

e. Was the coordination in the rear area complete?

f. Was the patrol order clear, complete, concise, and issued in a forceful, confident manner? Was the order tactically sound?

g. Were visual aids employed during issuance of the patrol order?

h. Were patrol members properly prepared, inspected, and rehearsed prior to the patrol's moving out?

i. Did the patrol pass through friendly units in the proper manner?

j. Did the patrol have security halt?

k. Was the formation suitable to terrain, cover, concealment, visibility, and proximity to known enemy positions?

l. Were the pace, point and compass men properly used?

m. Was the navigation accurate?

n. Was control maintained at all times?

o. Was security present and adequate?

p. Were subordinates properly utilized?

q. Were signals properly employed within the patrol?

r. Were rallying points designated?

s. Were rallying points easily distinguishable and tactically sound?

t. Were time elements and orders from higher headquarters adhered to?

u. Was the mission accomplished?

v. Was the location of the patrol known to the patrol leader at all times?

w. Did the patrol leave any visible signs along routes that would indicate their presence in the area?

x. Was all information obtained reported in the debriefing or patrol report?

y. Review the applicable portions of appendix VII, Patrol Tips. Did patrol leader, his subordinate leaders, or members of the patrol violate any of these tips.

APPENDIX XII

EXAMPLE PATROL ORDER (AMBUSH PATROL)

I. SITUATION

a. Enemy Forces. The enemy situation remains generally the same. The 85th Aggressor Division still occupies the high ground along the Tennessee Valley Divide Road; on the left is the 58th, in the center is the 116th, and on the right is the 174th Rifle Regiment. PW's and friendly air have confirmed that these units are in the process of being reinforced with replacements and equipment. Because of the reinforcement, they are presently able to occupy all of the critical terrain along the TVD. Ground between the FFL and this critical terrain is well covered with fire and patrols.

Strong point emplacements are well dug in and are generally composed of from 10 to 20 men.

Combat patrols from the 2d Ranger Bn raided and destroyed ---------------- bn-size CP installations at ---------------------, ---------------------, and -----------------. Resistance encountered indicated that the enemy in this area is well organized and morale is high.

The vehicle convoys being used to carry reinforcements to frontline units are using the following supply routes. NOTE: point out on map. Their size and composition range from three to six 2½-ton vehicles normally preceded by a scout ¼-ton. Rate of movement is slow due to the mountainous terrain. Scout ¼-ton precedes convoy from one to three minutes.

b. Friendly Forces. Friendly situation: The 10th Division remains in the same general area. On the left 1/16, in the center 1/26, and on the right 1/18. You will be passing through elements of the

---.

First Corps attacks tomorrow night at 2100 hours to seize the high ground along the TVD. The 10th Division, as part of First Corps, will participate in this attack.

c. Attachments. One battalion observer will be attached to each patrol and will join you here immediately following this briefing. With patrol one ----------------, with patrol two ----------------, with patrol three ----------------.

II. EXECUTION (figure 66)

a. Concept. You will depart friendly frontlines tonight and proceed through the enemy main battle line to a clandestine assembly

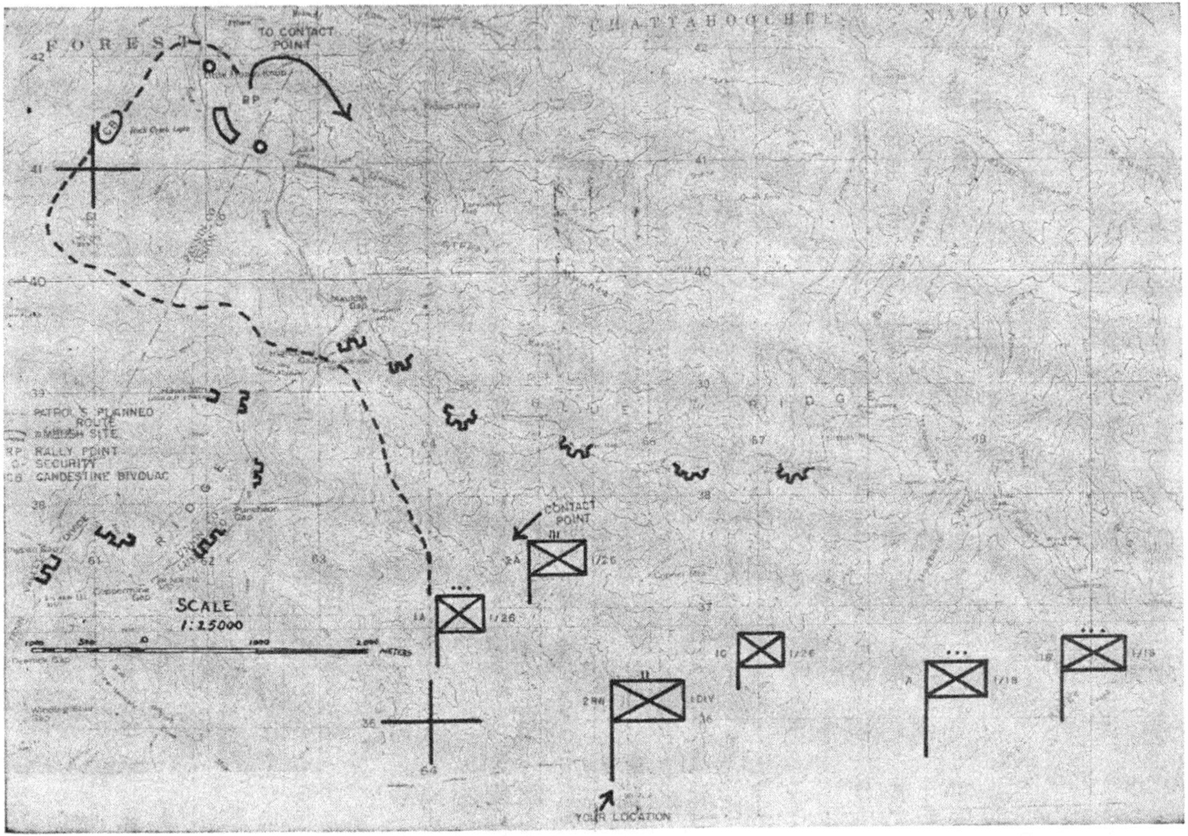

Figure 66. Map of combat area.

area of your own selection. Reconnoiter your objective area as necessary to select a suitable ambush site. Execute your ambush tomorrow night, then proceed to a designated rendezvous point where you will be met by friendly patrol.

b. Specific Duties. An overlay is available for each patrol indicating the OP through which you will pass, the time you will cross the LD, the area in which you will select your ambush site, and the location of the rendezvous point where you will meet the friendly patrol.

c. Coordinating Instructions. Vehicles depart at________________ hours for friendly frontlines where you will make coordination with your respective frontline unit. A guide will meet you at the detrucking point and guide you to a platoon CP. The unit commander will meet you at the CP to coordinate the passage of his frontlines. A representative of division artillery will be here at ________________ hours to coordinate supporting fires. A fire support overlay is available for each patrol. A light aircraft for aerial reconnaissance will be available at ________________ hours.

As soon as you have conducted your ambush, or if for any reason the ambush is not accomplished by 2400 hours tomorrow night, proceed to your rendezvous point where a friendly patrol will contact you between 0330 and 0500 hours________________________________.

I will debrief you at a point to which the contact patrol will guide you. If there is no contact because the attack was unsuccessful, or for any other reason, wait in the vicinity of the rendezvous point until darkness, then return to this area as a unit or by infiltration, as you deem appropriate. In any event, I will debrief you here upon your return.

III. ADMINISTRATION AND LOGISTICS

A hot meal will be served at ________________ hours. You will receive a hot meal on your return. Draw two meals of "C" rations here at ________________ hours. The battalion commander desires that no other rations be carried.

Turn in your equipment list to supply immediately after issuing your warning order. Draw equipment from supply at______ hours.

The aid station will remain initially in this area. You as patrol leader, will determine the disposition of all POWs and casualties.

IV. COMMAND AND SIGNAL

Each patrol will carry an AN/PRC–10; frequency ______________ megacycles; there is no alternate frequency. Call signs are as follows: NCS remains BACKBONE; fire support coordinator in the infantry command net is ASHCAN; patrol one is ARMFUL ONE; patrol two is ARMFUL TWO, etc. Check into the net at ________________ hours. Patrol radios will be turned on only for the following reasons: To check into the net, for emergency purposes, to request artillery support, and to render one of the following reports. Notify NCS when you have arrived in your clandestine assembly area. Use the code word PEPPER, then give the coordinates of your location in shackle. This will prevent friendly artillery from falling into your position. Notify NCS upon the completion of your ambush. If successful, use the code words CONDITION BLACK; if unsuccessful, use CONDITION BLUE. An AN/PRC–6 radio net may be opened at any time by your patrol. It has been determined that the enemy does not normally monitor low-power nets.

The shackle code for this operation is THE LAZY DOG; TANGO is zero and GOLF is nine.

The challenge and password from 1200 hours today until 1200 hours tomorrow is: FIRST TEAM. Challenge: FIRST. Password: TEAM. The challenge and password from 1200 hours ------------ -----------------------------: RED BUFFALO. The challenge and password from 1200 hours ---------------------------------- ------ until 1200 hours -- friendly patrol: BULL MARKET. A red star cluster will be fired by the friendly patrol as soon as they have arrived at the contact point.

APPENDIX XIII

EXAMPLE PATROL ORDER (GUERRILLA RAID)

I. INTRODUCTION

Good morning Rangers, I am ------------------, Bn S3 Liaison Officer. Alpha Company has been assigned an important mission and it looks as though you will have an opportunity to meet the aggressor before you leave for the southern front in Florida. Hold all questions until the completion of the briefing.

II. GENERAL SITUATION

United States forces are still containing the aggressor penetration in North Georgia and the amphibious landing to the south in Florida. However, the aggressor has infiltrated personnel from these areas and has organized a large guerrilla force within our rear area located in this general area (figure 67).

III. SPECIAL SITUATION

a. Enemy Forces.

1. The terrain in the area is wooded with scrub oak and tall pines. There are many open areas, particularly on hilltops. Along creeks and valleys, vegetation is thick and movement restricted. Most of the creeks are fordable.
2. The weather and light data are on this board which will be available to you after the briefing.
3. No identification has been made of the aggressor or guerrilla unit against whom you will operate.

NOTE: USE LARGE MAP TO SHOW.

4. The guerrilla unit controls the area bounded on the north by the UPATOI Creek and on the west by the OCHILLEE Creek. Several large airdrops of ammunition and equipment have been made by guerrilla aircraft. Intelligence indicates more of these materials are still near the areas used for drops. Enemy patrols have been seen along both creeks. The bridge crossing the UPATOI Creek is guarded.

NOTE: POINT OUT ON MAP.

5. The area between BUENA VISTA Road on the north and the UPATOI Creek on the south is the principal area of guerrilla activity. It is within this area that they have ambushed

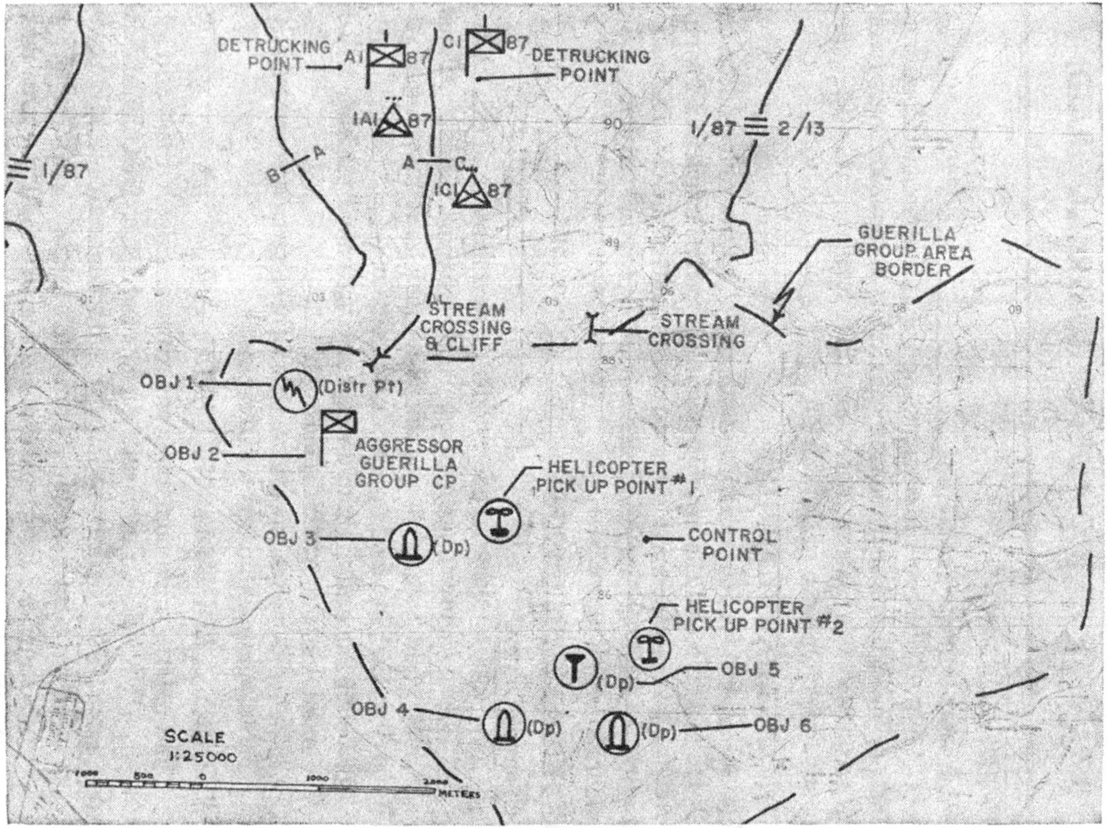

Figure 67. Map of combat area.

and destroyed our convoys. They do not control this area. The guerrilla forces have been particularly active along all roads. Also, these forces are known to have captured both friendly and civilian equipment and, on occasions, have used both against friendly troops.

6. Intelligence reports indicate that the aggressor force is comprised of two guerrilla bands of undisclosed strength. Each of these bands appears to have definite areas for stockpiling materials.

b. Friendly Forces.

1. First Ranger Battalion is continuing its action against the aggressor on the Florida front.
2. There will be no other friendly patrols operating in the area for the next 48 hours.
3. Upon completion of the mission, 1st BG, 87th Inf will conduct an attack to completely clear the guerrilla area.

c. Attachments.

1. Another Ranger battalion has recently been organized. One observer, who will be a cadreman in the new battalion, is to accompany each patrol and observe the patrol's method of operation.
2. The aircraft and pathfinders for this operation are attached to battalion; necessary coordination has been effected.

IV. MISSION

To destroy a series of ammunition and storage dumps at----------
--------within the area of guerrilla control by infiltration and stealth.

V. EXECUTION

a. General Plans. Patrols will depart this area by truck and proceed to the 1st BG, 87th Inf area; then proceed from company outposts of the 1st BG to assigned objectives; accomplish mission; rendezvous at designated landing zones for helicopter pickup and return by helicopter to our rear area here.

b. Specific Duties. Coordinates of friendly outposts and objectives are posted in your planning areas.

c. Coordinating Instructions.

1. All patrols will depart from our rear area at---------------- hours by vehicle. A guide will be available to guide the convoy to the area of the 1st BG, 87th Inf. At your detrucking points, a guide from the company through which you will pass will take you to the outpost from which you will depart. All patrols will clear the outposts not later than ------------ hours.
2. Time of return: Not later than ---------------- hours tomorrow.

3. Friendly units operating in this area have informed us of two stream crossing locations along the UPATOI Creek in areas lightly patrolled by the enemy. Arrangements have been made with partisans for a means of crossing the UPATOI Creek. The two stream crossing locations will be marked by a --------------------. Patrols -------------------- will cross at coordinates --------------------. Patrols-------- ----------will cross at coordinates-----------------------.

4. All patrols must accomplish their assigned missions and be at helicopter pickup points by ------------- hours. If helicopters are unable to participate in the operation, or if you do not reach the helicopter pickup point in time, your alternate route of return will be to infiltrate overland to coordinates --------------------.

5. Coordinates and helicopter pickup points are as follows:

Patrols 1 and 2	Patrols 3, 4, and 5
----------------------	------------------------------

6. The helicopter pickup points will be marked by pathfinder teams attached to the battalion.

7. Patrols will use radio call signs as verbal identification to the pathfinder teams at the pickup point. Contact pathfinder as soon as you reach your respective pickup points. Aircraft will not be contacted until all the patrols being lifted out of a pickup point have checked in.

8. No aerial reconnaissance will be available for this mission; however, aerial photos of portions of your area are available. You will find these in your planning area.

9. Aircraft will be spotted in the vicinity of ---------------- at --------------- hours for your loading rehearsal. At this time, a representative from the pathfinders will brief you on their support of the operation.

10. S2 representative will debrief you on your return at --------------------.
(Location)

VI. ADMINISTRATION AND LOGISTICS

a. No rations will be carried.

b. Turn in your equipment lists to the supply point located at ----- -------------------- upon completion of your warning order.

c. Draw your equipment and ammunition at ------------- hours.

d. All equipment drawn from the supply point will be cleaned and returned to the supply point immediately upon return to the rear area.

e. The handling of prisoners and wounded will be at the discretion of the patrol leader. --------------------------------------

f. Transportation from the rear area to the 1st BG, 87th Inf area will be spotted ------------------ at -------------- hours.

VII. COMMAND AND SIGNAL

a. Radios. AN/PRC-------------------- radios will be carried and will be for emergency use only. If an emergency exists and it is necessary to break radio silence, the phrase OPERATION RED will signify EMERGENCY EXISTS, NEED HELP. Radio the code word TOUCHDOWN after you have destroyed your objective.

b. Emergency.

1. Primary frequency
2. Alternate frequency

c. Call Signs.

S2 Liaison Officer------- UNDERWOOD
Radio Relay------------ UNDERWOOD WHISKEY
You will be------------- UNDERWOOD HAWK (your patrol No.)

My people will monitor both frequencies. There will be a battalion communications check at --------------------- at the front of ------ ---------------.

d. Challenge and Password. From 1200 hours today until 1200 hours tomorrow:

DEAD TIRED.

1200 hours today until 1200 hours following day:

GREEN WAGON.

Here is your time schedule.

NOTE: Refer to board.

Your observers and planning areas are:

The time is now -----------------.

What are your questions?

APPENDIX XIV

EXAMPLE LANE INSTRUCTOR'S INSTRUCTIONS (GUERRILLA RAID)

I. GENERAL

Problem No. 6 is a patrol-size raid with helicopter pickup against a series of guerrilla ammunition and storage dumps.

II. PURPOSE

a. The purpose of this problem is to acquaint the Soldier with the planning, preparation, and coordination necessary for the conduct of such an operation. It also serves to acquaint him with the procedures and sequence of events which will confront him during other complex operations.

b. Each lane instructor will make complete evaluation notes on each successive units leader's actions and order so as to present a comprehensive critique on completion of the problem. All patrol leaders will be given a verbal objective rating of their performance by use of the standard observation report.

III. CONDUCT OF LANE INSTRUCTORS (LI)

Each lane instructor will observe the actions and orders of his assigned patrol leader throughout the conduct of the problem. He will in no way interfere with the patrol leader in the latter's exercise of command. The lane instructor will "play the game" and only in the event of an emergency such as a serious accident, or if the Soldier leader becomes hopelessly lost, will he assume command or give specific directions to the Soldier leader.

IV. THE CONDUCT OF THIS PROBLEM

a. The Soldier will receive a briefing at 0800 hours of the first day of the problem.

b. After the briefing, Soldiers will move to planning areas for the preparation of the warning order, issuance of warning order, preparation of patrol order, drawing of equipment, issuance of patrol order, rehearsal, and inspection.

c. A critique will be conducted by the lane instructor on the first day after the patrol order, rehearsal, and inspection and on the second day after the termination of the problem.

d. The patrols will be moved by vehicle to designated outposts. Patrols will move to designated objectives for a coordinated attack, move to a rendezvous at helicopter pickup points, and be airlifted to the rear area.

e. Patrols will be debriefed at an area designated by the PI. LI's will be present at debriefing to ascertain correctness of information given by Soldiers.

V. ROTATION OF SOLDIER LEADERS

Phase I

a. Preparation and issuance of warning order.
b. Draw equipment.
c. Preparation and issuance of patrol order (critique by LI).

Phase II

a. Reorganization.
b. Rehearsal (critique by LI).
c. Inspection (critique by LI).
d. Movement to detrucking point.

Phase III

a. Reorganization.
b. Coordination at outpost.
c. Movement to UPATOI Creek.

Phase IV

a. Reorganization.
b. Movement to objective.
c. Conduct of assault.

Phase V

a. Reorganization at objective.

b. Movement to helicopter pickup point OCHILLE Creek and EIGHTH DIVISION Road.

c. Coordination at pickup point and conduct of movement to rear area.

d. Debriefing.

Lane instructors will tactically remove Soldier leaders at phases indicated above by declaring them injured or wounded by enemy action. At the point of relief, the old Soldier leader may pass on any instruction to the next in command, but only at that time. He then may function as a messenger or revert to role of rifleman at LI's discretion.

VI. LI'S DUTIES THROUGH DURATION ON PROBLEM AS FOLLOWS:

a. Briefing in classroom.	LI present to become familiar with briefing.
b. Patrols move to their planning area and prepare orders.	LI introduces himself to patrol in the planning area and directs time for warning order, gives instruction on his relationship to patrol and briefs on safety.
c. Patrols issue orders at designation.	LI present to hear and evaluate presentation of designated patrol leader. (Critiques all actions up to this point.)
d. Soldiers draw equipment.	LI's change into patrol uniform.
e. Patrol rehearsal.	LI's accompany their patrols and participate in rehearsal. (Critiques all actions up to this point.)
f. Patrol inspection.	LI's accompany their patrols to evaluate actions. (Critiques actions up to this point.)
g. Patrol proceeds with problem, to include coordination at outpost, crossing UPATOI Creek, movement to objective, assault of objective, movement to helicopter pickup point, helicopter pickup, airlift to rear area, and debriefing.	LI's accompany their patrols and remain with them through problem, to include debriefing.
h. Soldiers turn in equipment, mess, etc.	LI's prepare critique.
i. Critique.	LI's present in classroom for critique by PI—then move to planning areas for individual patrol critiques.

VII. SAFETY

a. LI's will enter the play of the problem only in the event of an emergency. At stream crossings and cliffs, LI will be safety officer for patrol.

b. LI's will caution Soldiers on use of grenade simulators and the firing of blanks.

c. Grenade simulators will not be placed in vehicles on objectives. Smoke grenades, simulating thermite grenades, will be placed on the ground at the objective.

d. In the event of an emergency, the lane instructor will utilize the AN/PRC–10 in contacting of the roving control points for assistance. In the event of radio failure, the injured Soldier will be moved to the nearest road and a fire built as an emergency signal.

e. Aid vehicles will be spotted at the control point located at bleachers at intersection of PINE TREE Road and FIRST DIVISION Road.

APPENDIX XV

EXAMPLE COMMAND POST INSTRUCTIONS

You are the CO or 1st Sgt of Co A, 1/87th Inf. Your job is to coordinate with the Ranger patrol leader whose patrol is passing through your positions. Do not volunteer any information to the patrol leader other than your name and organization, as this is a teaching point and we want the Soldiers to learn what to ask.

Use the mapboard provided to show the friendly and enemy situation, etc. If you do not know or are not permitted to supply an answer, state you do not know or cannot comply with the request.

Reverse your jacket so that your insignia does not show.

There are four Soldiers working in each lane with the following assignments:

1. Guide for detrucking area to your CP.
2. Officer or NCO at the CP.
3. Guide from the CP to the outpost.
4. NCO at outpost.

Instruct all guides to walk their routes in daylight and once after darkness to become familiar with them. The guide that brings patrol from the detrucking area should remain at the CP until the guide departs with the patrol from the CP. The detrucking guide then returns for the next patrol. As soon as OP guide takes his patrol to the outpost and guides them to the gap in the wire, he should return to the CP for the next patrol.

The following information should be given to the patrol leader if he asks:

1. You have no knowledge of enemy wire, boobytraps, mines, etc.

2. There are no friendly mines or boobytraps. You are not familiar with this terrain since your company moved into the position they now occupy two days ago. However, the terrain appears to be slightly hilly and the vegetation thick in places.

3. You have wire about 40 to 50 meters in front of your positions.

4. The only recent enemy contact has been an occasional probing patrol and several of your vehicles have been fired on along the roads. You know that the enemy is guarding the bridge across UPATOI Creek. You have observed enemy aircraft at night making what appeared to be an airdrop.

5. You cannot give the patrol any type of fire support.

6. You cannot support the patrol with a litter squad in front of your positions, but you will evacuate any wounded that are brought back to the outpost.

7. You have no navigational aids or signals that you can provide.

8. The patrol can have permission to designate their initial rally point to the rear of the outpost.

9. You will furnish a guide from your CP to your outpost.

10. You cannot provide the patrol with a spare radio. Your company frequency is --------------------.

11. The challenge and password is DEAD TIRED.

623989°—62——21

APPENDIX XVI

EXAMPLE OUTPOST INSTRUCTIONS

1. You are from 1st Platoon, Co. A, 1/87th Inf.

2. You have had no contact with the enemy. A patrol has been sighted this morning 50 meters to the left of your position.

3. You know of no enemy wire or mines to your front.

4. The terrain is slightly hilly and vegetation is thick in places.

5. Friendly wire is about 50 meters to the front. You will provide a guide to the gap.

6. There are no friendly mines or boobytraps in the area.

7. You cannot give a litter team but will evacuate all wounded if they are returned to your outpost.

8. You will be relieved about 0800 hours tomorrow. You came on duty at 0800 hours this morning.

9. You cannot give any support fires of any type.

10. If the patrol leader does not give the number of people in his patrol before he is finished with his coordination, ask him the number. Be sure the guide at your position counts the patrol through the gap in the wire.

11. You can give the patrol permission to have their initial rallying point near your PSN.

12. You do not have a spare radio for them to use.

13. You will pass on information to the person who relieves you.

14. Challenge and password is DEAD TIRED.

BOTH: DO NOT VOLUNTEER ANY OF THE ABOVE INFORMATION. WAIT UNTIL ASKED BY THE PATROL LEADER.

APPENDIX XVII

RANGER HISTORY

The history of the American Ranger is a long and colorful one and is a saga of courage, daring, and outstanding leadership. It is a story of men whose skills in the art of fighting have seldom been excelled.

The first Rangers were organized in 1756 by Robert Rogers (fig. 68), a native of New Hampshire, who recruited nine companies from among the Continentals. These units were identified as Rangers. Ranger techniques and methods of operation were an inherent characteristic of the frontiersmen in the American colonies prior to the American Revolution. It was Major Robert Rogers who first capitalized on such techniques and characteristics.

At that time, the British Army was engaged in fighting the French and Indian War (Seven Years' War). These Rangers were skilled in woodland warfare and were able to travel great distances over difficult terrain. As the chief scouting arm of the British, they were bold in procuring intelligence by scouting enemy forces and positions and taking prisoners. Roger's Rangers accompanied Wolfe's expedition against Quebec in The Montreal Campaign of 1760, and participated in the western campaign as far as Detroit and Shawneetown. They were sent by General Amherst to take possession of the northwestern posts, including Detroit. In the West, in 1763, Rogers and his men distinguished themselves in the Battle of Bloody Ridge.

Of interest, is the fact that Rogers' standing orders to his men (fig. 69), are still appropriate to today's Ranger type missions. Only the wording has changed.

The type of fighting used by these first Rangers was further developed during the Revolutionary War by Colonel Daniel Morgan, who organized a unit known as Morgan's Riflemen. These men, clad in frontiersman's buckskin garb, schooled in the Indian's methods of forest fighting, and armed with the deadly accurate frontiersman's rifle were without equal.

According to General Burgoyne, Morgan's men were ". . . the most famous corps of the Continental Army, all of them crack shots. . ."

Morgan's Riflemen fought at the Battle of Freeman's Farm (September 1777) and at the Battle of Cowpens (January 1781), where they inflicted heavy losses on the main body of British troops commanded by Colonel Tarleton. These successes were in large part due to the proper use of natural cover and surprise tactics.

Another famous Revolutionary War Ranger element was organized and led by Francis Marion. Marion's partisans, numbering anywhere from a handful to several hundred, operated both with and independently of other elements of General Washington's army. By disrupting British communications and preventing the organization

Figure 68. Major Robert Rogers.

Don't forget nothing.

Have your musket clean as a whistle, hatchet scoured, sixty rounds powder and ball, and be ready to march at a minute's warning.

When you're on the march, act the way you would if you was sneaking up on a deer. See the enemy first.

Tell the truth about what you see and what you do. There is an army depending on us for correct information. You can lie all you please when you tell other folks about the Rangers, but don't never lie to a Ranger or officer.

Don't never take a chance you don't have to.

When we're on the march we march single file, far enough apart so one shot can't go through two men.

If we strike swamps, or soft ground, we spread out abreast, so it's hard to track us.

When we march, we keep moving till dark, so as to give the enemy the least possible chance at us.

When we camp, half the party stays awake while the other half sleeps.

If we take prisoners, we keep 'em separate till we have had time to examine them, so they can't cook up a story between 'em.

Don't ever march home the same way. Take a different route so you won't be ambushed.

No matter whether we travel in big parties or little ones, each party has to keep a scout 20 yards ahead, twenty yards on each flank and twenty yards in the rear, so the main body can't be surprised and wiped out.

Every night you'll be told where to meet if surrounded by a superior force.

Don't sit down to eat without posting sentries.

Don't sleep beyond dawn. Dawn's when the French and Indians attack.

Don't cross a river by a regular ford.

If somebody's trailing you, make a circle, come back onto your own tracks, and ambush the folks that aim to ambush you.

Don't stand up when the enemy's coming against you. Kneel down, lie down, hide behind a tree.

Let the enemy come till he's almost close enough to touch. Then let him have it and jump out and finish him up with your hatchet.

Figure 69. Standing order, Rogers' Rangers, Major Robert Rogers—1756.

of loyalists to support the British cause, they contributed materially to Colonial victory.

Marion's group took part in the capture of Fort Johnson and in the victory of Charleston (1775). This victory gave the southern states a respite from fighting for nearly three years. Again active in 1780, Marion was instrumental in the capture of Fort Watson and Fort Motte, South Carolina, the following year. The loss of Fort Motte, on the line of communication between Camden and Charleston, was a great blow to the British cause. Marion's men also commanded the first line at the Battle of Eutaw Springs, taking many prisoners in one of the decisive battles of the Revolutionary War.

A favorite retreat of Marion's fighters was Snow's Island. Deep swamps bordered the island, and great quantities of game and livestock existed inland. Marion's men were able to launch sudden attacks from the island in any direction; surprising, killing, or capturing bands of Tories gathering to aid the British. After each action they would withdraw once again to the safety of the swamps.

The British Colonel Tarleton once pursued Marion's band through swamps and defiles for 25 miles. Arriving at a seemingly impassable swamp, Tarleton halted and cursed, ". . . the damned fox, the devil himself could not catch him . . ." Marion was to be known thereafter as the "Swamp Fox."

Marion's men were good riders and expert shots. They kept close watch on the British and detachments struck blow after blow, surprising and capturing small parties of Soldiers. They continually raided outposts, scouting parties, and lines of communication. There was no certain defense against Marion's guerrillas and their activity necessitated the presence of British regulars even in conquered regions. This organized partisan activity was most successful against an enemy of superior forces and discipline.

Marion's style of fighting was distasteful to the British commanders. It interfered with their plans for insuring and perpetuating their possession of the southern country which they sought to achieve by the establishment of military posts in different parts of North and South Carolina. Marion's rapid movements and secret expeditions cut off communication between posts and threw the whole system of government and military surveillance into confusion, aiding greatly in the Revolutionary cause.

The Civil War was again the occasion for the creation of special units such as Rangers. The Confederacy quickly capitalized on the advantages of this type of organization by authorizing the formation of partisan Ranger units. It was not until the summer of 1863 that the Union forces employed Ranger tactics and then only on a limited scale.

John S. Mosby, a master of the prompt and skillful use of cavalry, was one of the most outstanding Confederate Rangers. He believed that by resorting to aggressive action he could compel his enemies to guard a hundred points while he waited to attack any he chose.

The first real success of Mosby's Rangers was at Fairfax Courthouse, Virginia, located well behind Federal lines. Mosby had learned that enemy cavalry, infantry, and artillery units were there. He also knew the officer in charge was Colonel Percy Wyndham, a British Soldier of fortune fighting for the Union cause. Mosby's plan was bold: to infiltrate through the Federal lines and pluck the officer from the midst of the thousands of Soldiers protecting the roads west of Washington. His hope for success was based on the theory that to all appearances it was an impossibility.

Under cover of darkness, Mosby and 29 of the raiders infiltrated Federal outpost lines in the woods north of Centerville, Virginia. They cut the telegraph lines between Fairfax and Centerville preventing the sending of warning signals. The small band reached the outskirts of Fairfax early in the morning. As Mosby had hoped, the Federal headquarters was quite confident of its safety, positioned so far behind the lines. As a result, they did not employ a heavy sentry detail. Mosby and part of his command proceeded to a dwelling which was thought to be Wyndham's headquarters. It was the wrong house. In the meantime, Mosby learned that a Federal Soldier, whom they had captured, was one of the guards at the headquarters of Brigadier General Edwin H. Stoughton. Directing part of his detachment to Wyndham's quarters, Mosby himself took several men and set out to capture General Stoughton. Posing as Federal couriers, they gained entrance into the general's quarters and captured the sleeping officer. The detachment detailed to capture Wyndham reported that the Colonel had gone to Washington the afternoon before, however, they raided his quarters and captured the assistant adjutant general, a captain.

It was an unparalleled exploit. Twenty-nine men under a bold and aggressive leader had infiltrated through strong enemy lines to the very point where enemy officers slept, yanked them out of bed, laughed at their guards and disappeared before morning. They had captured a general, members of his staff, more than 100 other Soldiers and a large number of horses.

In March, 1863, Mosby defeated a much larger force of Federal troops near Chantilly, Virginia. When an attack which he had planned miscarried and his unit was pursued by a strong Federal cavalry unit, Mosby moved his men into a half-mile stretch of woods. From concealed positions, they delivered deadly carbine and pistol fire into the front and flanks of their pursuers killing five of them and wounding several others. One officer and 35 men, as well as a large number of horses, were captured. Not a Ranger was scratched!

At the Miskel Farm in the northern tip of Loudoun County, Virginia, Mosby and his band of 69 men were surprised by a force twice their size. During the bloody fight in the farmyard, Mosby rallied his men by shouting encouragement above the noise of the turmoil. His men heard, and delivered the stroke that brought victory. The results, as Mosby stated to Jeb Stuart, were ". . . nine of them killed—among them a captain and a lieutenant—and about fifteen too badly wounded for removal; in this lot, two lieutenants. We brought off eighty-two prisoners . . ."

Mosby men were mustered into the regular Confederate service for the remainder of the war. Initially they formed Company A, 43d

Battalion, Partisan Rangers. This unit was a part of the 1st Virginia Cavalry.

During the remainder of 1863, the Rangers were busy destroying wagon trains, capturing supplies, horses and mules, and obtaining information of Federal troop movements and dispositions. From May to July 1864, Mosby's men continued to plague Federal supply and ambulance trains capturing many Soldiers and large quantities of equipment. From one of the most successful of these raids, Mosby emerged with 200 prisoners, 500 to 600 horses, nearly 200 beef cattle, many valuable stores, and $112,000.00 in payrolls.

In the fall of 1864, an attempt by the Federal troops to build a railroad from Manassas Gap westward had to be abandoned because of Mosby's crippling raids.

The effectiveness of Mosby's operations is attested to by General Sheridan, who in his personal memoirs said, "During the entire campaign, I had been annoyed by guerrilla bands under such partisan chiefs as Mosby, White, Gilmore, McNeil, and others, and this had considerably depleted my line-of-battle strength, necessitating large escorts for my supply trains. The most redoubtable of these leaders was Mosby . . ."

Mosby was able to preserve and build up his organization over a two-and-a-half year period, within a few miles of the enemy capital. Numbered among his forces were men who knew practically every road and trail in Virginia and the location of the homes of many Confederate sympathizers behind the Federal lines. They struck in daylight and in darkness, whenever and wherever they could employ the element of surprise.

Mosby built his command to 800 before the end of the war, but the largest force he ever assembled for a raid scarcely exceeded 350. Usually, his forays were accomplished with a dozen to 80 men, because these small groups could more easily be concealed and moved about as necessity demanded.

Another prominent Ranger type unit was the cavalry squadron organized and led by General John Hunt Morgan. Morgan and his Confederate raiders began their famous attacks in December 1861. Their initial attack was on Lebanon, Kentucky, 60 miles from Morgan's Camp. During this raid they destroyed large quantities of stores and took several prisoners. A railroad bridge of military importance was burned, thus delaying the movement of Federal supplies to the front.

One of Morgan's most successful raids began in the summer of 1862. With his command of about 800 men, he left Knoxville and made his way westward to Sparta, Tennessee, encountering only a few scattered enemy along the way. Turning north at Sparta, Morgan crossed into Kentucky and captured a small garrison taking 400 prisoners and valuable stores including enough rifles to equip most of his unarmed men. The raiders then moved on to Glasgow and captured the garrison, where they destroyed more public stores. These two encounters were typical of the other raids Morgan conducted throughout his two-and-a-half week's march behind Union lines. During this time, he swelled his own ranks to 1,200 by recruiting en route, marched more than a thousand miles, captured seventeen towns, destroyed millions of dollars worth of Federal stores, dispersed many of the Home Guard, and raised Confederate morale to new heights. The losses to the Ranger force in both killed and wounded were less than 90.

Some insight as to the effectiveness of these raids is gained from General U.S. Grant's memoirs. Grant wrote, "Morgan was footloose and could operate where his information led him to believe he could do the greatest damage. During the time he operated in this way, he killed, wounded, and captured several times the number he ever had under his command at any one time. He destroyed many millions of dollar's worth of property in addition. Places he did not attack had to be guarded, as if threatened by him . . ."

The most famous raid of Morgan's Rangers started in July 1863. With a command of 2,400 men he attacked at Green River Bridge, Kentucky, but after a severe fight was forced to withdraw. Proceeding to Lebanon, Kentucky, they captured that garrison. Continuing to the Ohio River near Brandenburg, they crossed on two captured steamers after dispersing hostile troops on the far side. They encountered more militia at Corydon, Indiana, but quickly scattered them and captured the town. By this time, the whole countryside had risen in arms against them. Newspapers proclaimed an "Invasion of Indiana." Reinforcements were hurried in and gunboats on the river were rushed to intercept the Confederate marauders. Following a course roughly parallel to the Ohio River, bypassing Cincinnati, Morgan's men came to within a day's ride of Lake Erie—the deepest penetration of any Confederate force during the war. However, close on their heels was a Federal cavalry force. Near the end of July in the vicinity of East Liverpool, Ohio, Morgan was forced to surrender.

In spite of Morgan's surrender, the raid was successful. It drew off some of the forces which might have harassed Bragg's retreat from middle Tennessee and which might have helped Rosencrans at the Battle of Chickamauga later on. Colonel MacGowan, the Union Soldier, said, "Morgan's raid changed the whole aspect and results of military operations in Tennessee and Kentucky in the summer and fall of 1863. But for his diverting and delaying Burnside's movement upon Knoxville and the East Tennessee, Virginia, and Georgia railroad, that commander, with 28,000 men, would have joined Rosencrans three weeks before the Battle of Chickamauga was fought."

The cavalry arm was abolished in 1815 because of its cost. As time passed, the Indians, mounted on their war ponies, became a grave threat to expansion westward. The government organized a battalion of mounted Rangers to cope with this new problem. This battalion later expanded into the 1st Regiment Dragoons. It is interesting to note that their first mission under the leadership of General Leavenworth was a 500 mile march to the Upper Red River in Pawnee Territory.

The colonies, territories and early American states throughout our history formed Ranger units. They were activated to meet a crisis and deactivated for the most part immediately after the crisis had passed. The Connecticut, Texas (Thomas Knowlton's), and Arizona Rangers, to include the Mississippi Rifles, were some of the famous units.

With America's entry into the Second World War, Ranger units came forth once again to add to the pages of history. Major (later Brigadier General) William O. Darby organized and activated the 1st Ranger Battalion on 19 June 1942 at Carrickfergus, North Ireland. The volunteers were mostly from elements of the 1st Armored Division and the 34th Infantry Division. The members of this battalion were all handpicked volunteers. Six officers and 44 enlisted men of the battalion accompanied Commando troops in the Dieppe Raid on the northern coast of France. These men learned much of the German's fighting methods and defenses, which proved of inestimable value to the Rangers in later operations. The 1st Ranger Battalion participated in the initial North Africa landing at Arzeu, Algeria, and in the Tunisian battles, where they executed a number of hazardous night attacks over difficult and treacherous terrain. The battalion was awarded the presidential Unit Citation for distinguished action which included operations in the critical battle of El Guettar.

The 3d and 4th Ranger Battalions were activated and trained by Darby in Africa near the close of the Tunisian Campaign. These three battalions made up what was known as the Ranger Force.

Darby's Ranger battalions spearheaded the Seventh Army landing at Gela and Licata during the Sicilian invasion and played a role in the subsequent campaign which culminated in the capture of Messina. In the Salerno engagement on the Italian peninsula, the Ranger force fought for 18 days to hold Chiunzi Pass against eight German counterattacks in the Venafro Battles. The Rangers experienced fierce winter and mountain fighting in clearing the entrance to the narrow pass leading to Cassino. At Anzio, they had the mission of overcoming beach defenses, clearing the town, and forming a defensive perimeter.

On the night of 30 January 1944, the 1st, 3d, and 4th Ranger Battalions launched an attack on Cisterna. The Rangers were annihilated at Cisterna. The surviving original Rangers who had volunteered in Ireland, Scotland or at Arzeu, Algeria, were returned to the United States after this action. The newer members were transferred to the American-Canadian Special Service Force which was engaged in holding the lower stretches of the Mussolini Canal facing Littoria and which shortly thereafter joined in the march on Rome.

The Special Service Force, like the Rangers, was a highly trained volunteer unit that specialized in night raiding and beach landings. It had led the American drive into Kiska in the Aleutians. It participated in the drive to the Gustav Line in Italy clearing the Mount Majo hills. Following this action, it moved to Anzio taking up positions on the right flank along the Mussolini Canal and participated in the breakout to Rome. In operation ANVIL, it spearheaded the landing in Southern France and fought with the Seventh Army near Belfort Gap.

The 2d and 5th Ranger Battalions participated in the D-day landings (6 June 1944) on Omaha Beach, Normandy. Attached to the 116th Infantry, 29th Division, Companies D, E, and F of the 2d Ranger Battalion accomplished their mission of capturing Pointe du Hoe, a German coastal battery. The two battalions then assisted in the capture of Grandcamp and the mopup of scattered enemy opposition Grandcamp and Isigny.

The 5th Ranger Battalion participated in operations in the Bay of Brest area. Operating on the left flank, they assaulted and captured three of the numerous defenses which extended the seven miles to Recouvrance.

Later in September 1944, the 2d Battalion, attached to Task Force Sugar of the 29th Infantry Division, drove through numerous outpost strongpoints to reach the German main line of resistance. The Le Conquet Peninsula was the next objective to Task Force Sugar. The 2d Battalion assisted in this by breaking into the 280-mm gun positions (batteries Graf Spee) and forced the surrender of Le Conquet garrison commander and 814 men. The 5th Battalion met little opposition in the reduction of the Le Conquet peninsula defense.

During the Rhineland Campaign, 6–8 December 1944, the 2d Ranger Battalion, operating in the Hurtgen Forest, captured critical heights near Bergstein creating a salient in the German lines. Although counterattacked five times and subjected to continuous artillery fire, the unit held the ground which offered observation of the key town of Schmidt, as well as of the Roer river dams. The salient created by the attack reached the most easterly point to which the Allies had driven.

In November 1944, General Patton assigned the 5th Ranger Battalion to XX Corps. A force consisting of the 6th Cavalry Group and the Ranger battalion had the mission of screening the XX Corps southern flank.

In February and March 1945, the 5th Ranger Battalion, while attached to the 94th Infantry Division, accomplished a mission of great consequence to the success of the Allied operations in the Saar river area. Under cover of darkness, the battalion infiltrated the enemy frontline positions and seized the high ground commanding the main German military supply route west of Zerf. Two counterattacks were repulsed and after five days of fighting the 5th Ranger Battalion had killed 378 men, wounded an estimated 550, captured 562 more, and destroyed two armored vehicles. Seizure of their assigned objective aided the armored breakthrough which overran Trier and brought elements of the XX Corps to the banks of the Rhine river.

The 6th Ranger Battalion, operating in the Pacific, was the only Ranger unit fortunate enough to have been assigned only those missions applicable for Rangers. All of its missions, usually of task force, company or platoon size, were behind enemy lines, involved long-range reconnaissance and hard-hitting long-range combat patrols. The three most noteworthy were during the campaigns in the Philippines.

The first American contingent to return to the Philippines was the 6th Ranger Battalion with the mission of knocking out the coastal defense guns, radio stations, radar stations, and other means of defense and communications in Leyte harbor. On A-day minus three, the 6th Ranger Battalion was landed from fast attack-type converted destroyers, in the midst of a storm, on Dinagat, Suluan, and Homonohan islands in Leyte bay. Their mission was successfully accomplished with hours to spare.

Later, a reinforced company from the 6th Battalion formed the entire rescue force which liberated American and Allied prisoners of war from the Japanese Prison Camp at Cabanatuan, the Philippines in January 1945. They made a 29-mile forced march into enemy territory, obtained full support of local civilians and guerrillas, and determined accurately the enemy's dispositions. They crawled nearly a mile through flat and open terrain to assault positions, destroyed a Japanese garrison nearly double the size of the attacking force, and in the dark, assembled over 500 prisoners of war. The prisoners were evacuated from the stockade area within twenty minutes after the assault began. In this action, more than 200 enemy troops were killed. Ranger losses were two killed and ten wounded.

The Ranger's last mission was the 250 mile trek behind enemy lines, by B Company to the city of Aparri on the northern tip of Luzon. Aparri was the last seaport and major city held by the Japanese forces. For twenty-eight days behind the lines, they successfully infiltrated and reconnoitered the Japanese defenses at Aparri. They prepared the landing facilities at Camalugian Airfield for the 11th Airborne to make one of the major airdrops of the Pacific Campaign. Following the successful airdrop the Rangers initially supplied the point and later the flank security for the 11th Airborne Task Force driving southward along the Cagayan river to link up with the 32d Infantry Division and thus end the Philippine Campaign. It is noteworthy that all of the Japanese prisoners captured during this operation and turned over to the 11th Airborne Division were captured by one platoon from the 6th Ranger Battalion.

Another Ranger type unit in the Pacific was the 5307th Composite Unit (Provisional), organized and trained as a long-range penetration unit for employment behind enemy lines in Japanese-held Burma. Commanded by Brigadier General (later Major General) Frank D. Merrill, its 2,997 officers and men became popularly known as "Merrill's Marauders." From February to May 1944, the operations of the Marauders were closely coordinated with those of the Chinese 22d and 38th Divisions in a drive to recover northern Burma and clear the way for the construction of the Ledo road which was to link the Indian railhead at Ledo with the old Burma road to China. The Marauders were foot-Soldiers who marched and fought through jungles and over mountains from the Hukawng Valley in northwestern Burma to Myitkyina on the Irrawaddy river. In five major and 30 minor engagements, they met and defeated the veteran Soldiers of the Japanese 18th Division. Operating in the rear of the main forces of the Japanese, they prepared the way for the southward advance of the Chinese by disorganizing supply lines and communications. The climax of the Marauders' operations was the capture of the Myitkyina airfield, the only all-weather strip in northern Burma. This was the final victory of "Merrill's Marauders" which was disbanded in August 1944.

The men composing "Merrill's Marauders" were volunteers from the 33d Infantry Regiment, the 14th Infantry Regiment, the 5th Infantry Regiment and from infantry regiments engaged in combat in the southwest and South Pacific. These men responded to a call from the Chief of Staff, General George C. Marshall, for volunteers for a hazardous mission. These volunteers were to be of a high state of physical ruggedness and stamina and to be from jungle-trained and jungle-tested troops.

Prior to their entry into the Northern Burma Campaign, "Merrill's Marauders" trained in India under the overall supervision of Major General Orde C. Wingate, British Army. Here they were trained in long-range penetration tactics and techniques of the type developed and first employed by General Wingate in the operations of the 77th Indian Infantry Brigade in Burma from February to June 1943.

With the outbreak of hostilities in Korea in June 1950, the need arose once again for Rangers. On 25 August 1950 at Camp Drake, Japan, the 8213th Army Unit was organized from volunteers in the Far East. The 8213th was referred to more informally as the Eighth Army Ranger Company and was attached to the 25th Infantry Division. It participated in the "drive to the Yalu" and was deactivated in March 1951.

Fourteen airborne Ranger companies were formed and trained at the Ranger Training Command, Fort Benning, Georgia, between September 1950 and September 1951.

The 1st, 2d, 3d, 4th, 5th, and 8th Ranger Infantry Companies (Airborne) were assigned to divisions throughout the Eighth Army in Korea and used as line infantry. These units were deactivated in September 1951 and their highly trained personnel were spread throughout the army.

In October 1951, the Chief of Staff, General J. Lawton Collins, directed that Ranger training be extended to all combat units in the army in order to develop the capability of carrying out Ranger type missions in all infantry units of the army. The Commandant of the Infantry School was directed to establish a Ranger Department for the purpose of conducting a Ranger course of instruction. The overall objective of Ranger training was to raise the standard of training within combat units. This was a twofold mission: first for United States Army Infantry School to train a Ranger cadre, and second for infantry units to conduct Ranger training. The goal was to provide one Ranger qualified officer per rifle company and one noncommissioned officer per rifle platoon.

In July 1954, General Mathew B. Ridgeway gave additional emphasis to the Ranger program when he made it mandatory that all newly commissioned Regular Army officers of Infantry, Armor, Artillery, Corps of Engineers, and Signal Corps select and attend Airborne or Ranger training. In June 1955, Regular Army second lieutenants of the Military Police Corps were included. Some 70 percent of the Regular Army lieutenants take both Airborne and Ranger training.

Ranger training is a concept directed by the Chief of Staff, United States Army, to raise the level of infantry training, army-wide. Over the years since October 1951, this concept of Ranger training has developed into the present Ranger program. This program is built upon what has been inherited from the Rangers of the past. In World War II, the concept of employment was Ranger battalions. During Korea, the concept was Ranger-Airborne companies. Now, it is individual Ranger training; the highest form of individual infantry training in the army today. All Rangers have a common ability to operate over varied terrain at night. The Ranger today is an individual highly imbued with Ranger esprit and drive, who returns to his unit to integrate his training into present training programs.

Since March 1952, when the first Ranger class under the current program completed training, some 8,000 United States officers and enlisted men have earned the coveted Ranger Arc. These graduates provide the nucleus to cadre special trained units requiring Ranger characteristics and skills, should The Department of the Army decide to organize such units at Corps or Army level. Ranger training is also applicable to portions of Special Forces operations.

The Ranger concept has been carried to various parts of the world, over 285 Allied officers, representing 34 countries have completed the Ranger course. In addition some countries such as Viet Nam, Iran, Japan, West Germany, and Nationalist China have established their own Ranger courses patterned after the United States version.

GLOSSARY

Abatis—Obstacle made of cut or fallen trees or branches, or small trees or saplings bent down and frequently reinforced with barbed wire. A dead abatis is made of cut or fallen trees or branches; a live abatis of trees or saplings bent down.

Ambush—A surprise attack from a concealed position upon an unsuspecting moving or temporarily halted enemy.

Area or zone reconnaissance—The gathering of information within a defined area. Reconnaissance by fire may be a technique used in accomplishing this type mission.

Bear claws—An arrangement consisting of a metal ring with individual hooks attached in a manner similar to the claws of a bear, used for securing rope in the negotiation of mountain or cliff obstacles.

British crawl—A method of crawling on top of a rope by lying on the chest with one leg and foot hooked over the rope and letting the other leg hang down, pulling with the hands.

Buddy team—The smallest unit designation in Ranger training; two Rangers assigned together for the purpose of obtaining the natural support that is necessary to overcome the many obstacles and pitfalls of Ranger training or in combat.

Cavitation—Term referring to the formation of an air pocket that is sucked down by the propeller. The air pocket forces the water away from the propeller causing it to spin more rapidly and the motor to race.

Clandestine assembly area—An assembly area located in terrain not protected by friendly forces where a patrol can halt for a short period to plan, reorganize, and rest prior to continuing on its mission.

Cliffhead team—Section of a patrol in a cliff assault assigned the task of securing the cliffhead.

Counterforce—When two demolition charges are placed exactly opposite each and flush with the target, and the detonation is simultaneous. This arrangement is called a counterforce setup.

Diamond charge—A demolition designed for cutting a steel rod or shaft. This charge has a diamond shape pattern.

Grape shot—Is a makeshift demolition made by placing an inverted funnel in the bottom of a No. 10 can, fill funnel with pieces of metal and fill container with explosives, priming in the exact center rear; used against ground troops.

Grappling hooks—Hooks fixed to the end of a rope or line so that rope or line may be attached to an object by throwing and catching the hook—used in cliff assault.

Grenade necklace—Hand grenades connected by detonating cord and twine. Normally used in setting up an ambush killing zone.

Helicopter rappel—A method by which a man can leave a hovering helicopter by the use of a rappel seat and rappelling rope.

Intelligence—Evaluated and interpreted information covering an actual or possible enemy area or operation.

Lane—A lane (as used in this text) is an area of land on which Ranger patrol problems are run. This land area is large enough to allow flexibility in the play of the problem for various patrol action. The area of maneuvers for a single patrol.

Lane instructor—Experienced cadre of a Ranger training program who accompanies the Soldier on the simulated combat exercises. The instructor term signifies that the cadre may correct erroneous decisions when the student-Soldier makes them and may halt the patrol to cover a vital teaching point. This system is used with inexperienced personnel. Its primary disadvantage is that Soldiers may attempt to rely upon the instructor and fail to develop the ability to make decisions on their own.

Linear shaped charge—A demolition charge used for cutting steel plate and beams.

Loading manifest—Compilation of the cargo and/or personnel loaded on a vessel, aircraft or vehicle.

Monroe charge—A principle used with the shaped charge which causes all the forces to be placed in one small area.

Offset—An "offset" is a planned magnetic deviation to the right or left of the straight line azimuth to an objective. It is used to verify your exact location in relation to the objective.

Route reconnaissance—The gathering of information about a specific route or routes. Mobility often favors the use of air or vehicular transportation.

Patrol principle—Established doctrine relative to patrolling operations, e.g., continuous security is a patrolling principle. Security halts and use of point man are techniques used to facilitate this principle.

Patrol technique—Method of performing acts, especially the detailed methods used by patrol leaders in performing their assigned tasks. Refers to the basic methods of using equipment and personnel. Patrol techniques are derivative or means to effect patrolling principles. An established technique may be overlooked due to necessity; however, patrolling principles are never violated.

Platter charge—Pressure and Breaching—A type of demolition used on concrete bridges and abutments.

Point reconnaissance—Reconnaissance of a specific location. Foot patrols within the objective area normally offer the greatest degree of success if the point is occupied by the enemy.

Rallying point—A rallying point is a place where a patrol can assemble and reorganize if unavoidably separated or dispersed. A rallying point should offer cover and concealment, be defensible for at least a short time, and be easily recognized and known to all patrol members.

Ranger operations—Overt operations by highly trained infantry units to any depth into enemy held areas for the purpose of reconnaissance, raids, and general disruption of enemy operations. Depth and duration of the operation is limited only by resources for delivery of the forces and their mission.

Ranger type training—Ranger training is realistic, rough, and to some degree, hazardous. It consists of a minimum of academic instruction. Training is designed to develop the individual's self-confidence, leadership, ability to command, and skill in the application of basic infantry techniques.

Rendezvous point—In patrolling a predetermined point at which patrol members meet other members of a patrol or other personnel such as partisans or agents.

Ribbon charge—A demolition used to cut beams. This charge is long, thin, and narrow or "ribbon like" and the point of detonation is on the end.

Sand Delta—A sand flat caused by flowing water at a point where two bodies of water come together.

Scaling ladder—A ladder used for getting over an obstacle, operation used in cliff assault.

Security halt—After passing through a friendly position at night, move a short distance, halt and listen. Every man should try to detect the enemy. Have security halt just far enough from the friendly position to be safe from mortar and artillery fire that may fall on it if the enemy has learned its location. The security halt should be short.

Side hill cuts—This type of roadblock is placed with demolitions by the employing unit. In this type of roadblock consider bypass routes by the enemy.

Terminal guidance of aircraft—Ground controlled guidance at a landing site when the aircraft is used in a normal manner.

Toggle ropes—A rope with with a loop on one end and a wooden or metal T-peg on the other used for utility or cliff work. Easily joined to form a longer rope.

Tyrolean traverse—A method used in mountaineering to make a lateral movement by the use of rope.

INDEX

	Paragraph	Page
A-frame, use	app. V	233
Aggressor:		
Areas	14	16
Control officer	18	23
Aids for rope installations	app. V	233
Air:		
Guard in boat	60	121
Safety	30	50
Airmobile operations	32	57
Ambush:		
Conduct	39	63
Defense against	40	65
Patrol problem No. 3	app. IV	208
Techniques	app. III	160
AM, FM, siting radio sets	55	96
Anchors for rope installations	app. V	233
Antennas, Radio	55	96
Antiguerrilla:		
Missions	64	135
Operations	61	130
Areas:		
Aggressor	14	16
Clandestine	54	90
Destruction	63	133
Geographical training	51	83
Reserve	14	16
Artillery, Coordination	59	14
Assault:		
Boats	59	114
Cliff techniques	8, 42, app. III	7, 69, 160
Attack against guerrillas	66	136
Bayonet, objective	app. III	160
Beach landing	44	70
Belays, use	app. V	233
Blisters, treatment	56	101
"Blood-Poisoning", definition	56	101
Boats:		
General	59	114
Selection	60	121
Brain injuries	56	101
Bridges, rope	app. V	233
Camouflage measures	39	63
Checklists:		
Equipment	26	27

	Paragraph	Page
General critique	app. XI	298
Use	24	25
Chest wounds	56	101
Clandestine assembly areas	54	90
Cleanliness necessity	56	101
Cliff evacuation	app. V	233
Cliffhead team, withdrawal	45	76
Climbers:		
Balance climbing	app. V	233
Duties	44	70
Clothing, extended operations	53	88
Combat, hand-to-hand	9, app. III	9, 160
Command post:		
Forward company	14	16
Instructions example	app. XV	318
Commander:		
March considerations	51	83
Medical kit items	56	101
Communications:		
Ambush	39	63
Methods	23	25
Radio	55	96
Compass, night course	app. III	160
Concept, training	5	6
Critique:		
General checklist	app. XI	298
Phases	app. IV	208
Purpose	24	25
Debarkation to beach	59	114
Demolitions, acquaintance with	app. III	160
Desert:		
Operations	37	60
Training areas	9	9
Disease, communicable	56	101
Encirclement of guerrillas	66	136
Enemy:		
Representation	16	21
Situation during instructions	app. III	160
Wire	app. IV	208
Equipment:		
Cliff assaults	42	69
Extended operations	53	88
TOE	29	35
Escape and evasion methods	app. III	160
Evacuation:		
Cliff	app. V	233
Plan	13	16
Report	29	35
System	28	30
Exercises, runs, and games, general	app. III	160
Eye wounds	56	101

	Paragraph	Page
Fire:		
Plan coordination	39	63
Support	59	114
Footwear, considerations	51	83
Formations, combat	app. III	160
Formula for length wave sections	55	96
Grading, evaluation system	25	26
Guerrilla:		
Characteristics	62	133
Clandestine base	54	90
Combating the	63	133
Operations against	33	58
Raid (patrol order)	app. XIII	309
Handling lines, vertical	app. V	233
Hand-to-hand combat	app. III	160
Hemorrhage general	56	101
History, Ranger	app. XVII	321
Holds, climbing	app. V	233
Inspections, purpose	app. III	160
Infiltration, operations against	33	58
Injuries:		
Brain	56	101
Cold weather	56	101
Intelligence:		
Guerrilla plans	63	133
Importance	app. III	160
Items for student problems	29	35
Jaw wounds	56	101
Jungle:		
Operations	37	60
Training areas	9	9
Knots, types	app. V	233
Lane instructor (LI):		
General	17	22
Guerrilla raid instructions	app. XIV	314
Problem notes	app. IV	208
Water responsibilities	30	50
Layout, clandestine area	54	90
Leaders, rotation of Soldier	app. XIV	314
Logistics, extended patrol unit	57	113
Map reading	8, 28, app. III	7, 30, 160
March security	40	65
Marshalling actions	59	114
Medical:		
Aid kit	56	101
Considerations	56	101
Shock	56	101
"Merrill's Marauders", history	app. XVII	321
Missions, antiguerrilla	64	135
Morphine, use	56	101

	Paragraph	Page
Mountain:		
Features	9	9
Operations	37	60
Techniques	app. III	160
Walking	app. V	233
Mountaineering, lesson outlines	app. V	233
Naval fire support	59	114
Navigation:		
Critique checklist	app. XI	298
River techniques	58, 60	114, 121
Northern operations	37	60
Nuclear warfare:		
Application	1	3
Preparation	46	80
Obstacles:		
River	60	121
Types	41, app. III	68, 160
Operations:		
Against infiltration	33	58
Antiguerrilla	61, 65	130, 135
Cliff assault	42	69
Extended.	47	81
Mop-up	49	81
Night	34	58
Signals, waterborne	59	114
Special areas	37	60
Waterborne	58	114
Orders:		
Ambush patrol	app. XII	305
Guerrilla raid	app. XIII	309
Major Robert Rogers'	app. XVII	321
Patrol planning techniques	8, app. III	7, 160
Orientation, control personnel	12	15
Outpost instructions example	app. XVI	320
Outlines, lesson subjects	app. III	160
Patrol:		
Ambush order	app. XII	305
Ambush problem No. 3	app. IV	208
Antiguerrilla operations	66	136
Combat checklist	26	27
Conduct	app. VIII	285
Daily plan	app. VIII	285
Delivery to target area	50	83
Leader's duties	25	26
Logistics	57	113
Night combat problem No. 2	app. IV	208
Order format	app. VIII	285
Planning	app. III	160
Reconnaissance checklist	26	27
Report (fig. 44)	17	22
Tips	app. VII	278
Patrolling, considerations	36, app. VIII	59, 285

	Paragraph	Page
Personnel:		
Orientation	12	15
Radio	55	96
Safety	20	24
Selection	9, 51	9, 83
Supply	21	24
Photograph reading	app. III	160
Physical conditioning, objective	app. III	160
Pneumatic raft, use of	app. III	160
Points:		
Detrucking	14	16
Rallying	app. XI	298
Poncho raft, use	app. III	160
Principal Instructor (PI), general	18	23
Prisoner, seizure	49, app. III	81, 160
Problems:		
Aggressor radio net	app. X	295
Ambush (No. 3)	app. IV	208
Communications support	app. X	295
Holding enemy installation (No. 4)	app. IV	208
Logistics	app. IV	208
Night combat patrol (No. 2)	app. IV	208
Obtaining prisoners (No. 5)	app. IV	208
Outline to training memorandum	app. IX	291
Patrol walk through (No. 9)	app. IV	208
Radio net	app. X	295
Raiding guerrilla forces (No. 6)	app. IV	208
Reconnaissance (No. 1)	app. IV	208
Purpose, Ranger training and Ranger operations	1	3
Radio:		
Aggressor net	app. X	295
Communications	55	96
Interpatrol net	app. X	295
Problem net	app. X	295
Rafts, Pneumatic and poncho	app. III	160
Raid:		
Landings	44	70
Planning	35	59
Rallying point:		
Characteristics	app. III	160
In problems	app. IV	208
Ranger history	app. XVII	321
Rapelling, use	app. V	233
Reading, Map and photograph	app. III	160
Reconnaissance:		
Clandestine areas	54	90
General	9	9
Specific location	49	81
References pertinent to Ranger training	app. I	150
Rehearsal:		
At sea	59	114
Conduct training program	9	9
Rendezvous, all elements	59	114
Reorganization, all elements	59	114

	Paragraph	Page
Reports:		
Patrol	17	22
Spot	28	30
Resupply:		
Dispersed units	49	81
By air	57	113
"Rip, Tear, Grub, and Dig", meaning	app. VI	268
River:		
Crossing expedients	app. IV	208
Navigation	60	121
Robert Rogers, Major	app. VII	278
Roadblock:		
Establishment	41	68
Techniques	app. III	160
Rope:		
And knots	app. V	233
Bridges	app. V	233
Fixed	app. V	233
Safety	30	50
Safety:		
Aidmen, ambulance drivers	20	24
Air	30	50
Rope	30	50
Training personnel	30	50
Water	30	50
Schedules, training	app. II	152
Secrecy of movement	60	121
Security:		
Clandestine area.	54	90
March	40	65
Zones	54	90
Shock, definition	56	101
Signals, waterborne operations	59	114
Stay-behind unit	49	81
Stealth, purpose	66	136
Stream crossing SOP	30	50
Supply personnel	21	24
Survival:		
Lesson outline	app. VI	268
Methods	8	7
Objective	app. III	160
Techniques, river patrolling	60	121
Terrain preparation	14	16
Tests, confidence	app. III	160
Training:		
Areas	9	9
Concept	5	4
Foot movement	51	83
Memorandum (problem outline)	app. IX	291
Phases	6	5
Safety	30	50
Schedules	9, app. II	9, 152
Subjects	8	7
Vehicle mobility	51	83

	Paragraph	Page
Transportation:		
General	22	25
To docks or on-shore	59	114
Traverse, suspension	app. V	233
Tributaries reconnoitered	60	121
Troops:		
Aggressors	9	9
Friendly	15	21
Orientation	11	15
Unit at sea, actions	59	114
Vehicle mobility training	51	83
Walking, mountain	app. V	233
Warning order	app. III	160
Water safety	30	50
Wire, enemy	app. IV	208
Weapons:		
Requisition	29	35
Selection	52	88
Training	51	83
Withdrawal:		
Patrol from target area	50	83
Raid party	45	76
Wounded, treatment	app. III	160
Wounds:		
Care	56	101
Eye	56	101
Zones, security system	54	90

BY ORDER OF THE SECRETARY OF THE ARMY:

G. H. DECKER,
General, United States Army,
Chief of Staff.

Official:

J. C. LAMBERT,
Major General, United States Army,
The Adjutant General.

Distribution:

Active Army:

DCSPER (2)
ACSI (2)
DCSOPS (2)
DCSLOG (2)
ACSRC (2)
CRD (1)
COA (1)
CINFO (1)
TIG (1)
TJAG (1)
TPMG (1)
Tech Stf, DA (1)
USCONARC (5)
ARADCOM (2)
ARADCOM Rgn (1)
OS Maj Comd (5)
LOGCOMD (1)
MDW (5)
Armies (5)
Corps (3)
Div (2)
Bde (1)
Regt/Gp/Bg (1)
Bn/Sqd (1)
Co/Btry/Trp (5)
Tng Comd (25)
Instl (1)
Br Svc Sch (5) except
USAIS (371)
USARMS (35)
USAES (20)
MFSS (15)
USAAVNS (6)
USACGSC (10)
USAINTC (9)
USA Jungle Warfare TC (150)
USA Cold Weather & Mt Sch (5)
Units organized under following TOE: 17-22 (1)

NG: Corps Arty, Div, Div Arty, Bde, Gp, Regt, BG, Combat Command, Inf Bn, Engr Bn (2); Units organized under following TOE's: 7, 17, 33-106, 33-107 (3).

USAR: Same as Active Army except allowance is one copy to each unit.

For explanation of abbreviations used, see AR 320-50.

U.S. GOVERNMENT PRINTING OFFICE: 1962

Made in the USA
Coppell, TX
18 October 2026

93133429R00204